Bergson's Scientific Metaphysics

Also available from Bloomsbury

Advances in Experimental Philosophy of Science, edited by Daniel A. Wilkenfeld and Richard Samuels
Biopolitics After Neuroscience, Jeffrey P. Bishop, M. Therese Lysaught, Andrew A. Michel
Material Objects in Confucian and Aristotelian Metaphysics, James Dominic Rooney
The Metaphysics of Contingency, Ferenc Huoranszki

Bergson's Scientific Metaphysics

Matter and Memory Today

Edited by

Yasushi Hirai

BLOOMSBURY ACADEMIC
LONDON • NEW YORK • OXFORD • NEW DELHI • SYDNEY

BLOOMSBURY ACADEMIC
Bloomsbury Publishing Plc
50 Bedford Square, London, WC1B 3DP, UK
1385 Broadway, New York, NY 10018, USA
29 Earlsfort Terrace, Dublin 2, Ireland

BLOOMSBURY, BLOOMSBURY ACADEMIC and the Diana logo are trademarks
of Bloomsbury Publishing Plc

First published in Great Britain 2023
This paperback edition published 2024

Copyright © Yasushi Hirai and Contributors 2023

Yasushi Hirai has asserted his right under the Copyright, Designs and Patents Act, 1988, to be identified as Editor of this work.

For legal purposes the Acknowledgements on p. xiii constitute an extension of this copyright page.

Series design by Charlotte Daniels
Cover image: Soft wavy stripes flow in orange, purple and red with selective focus on the center (© Flavio Coelho / Getty Images)

All rights reserved. No part of this publication may be reproduced or transmitted in any form or by any means, electronic or mechanical, including photocopying, recording, or any information storage or retrieval system, without prior permission in writing from the publishers.

Bloomsbury Publishing Plc does not have any control over, or responsibility for, any third-party websites referred to or in this book. All internet addresses given in this book were correct at the time of going to press. The author and publisher regret any inconvenience caused if addresses have changed or sites have ceased to exist, but can accept no responsibility for any such changes.

A catalogue record for this book is available from the British Library.

A catalog record for this book is available from the Library of Congress.

Library of Congress Control Number: 2023937565.

ISBN:	HB:	978-1-3503-4197-5
	PB:	978-1-3503-4201-9
	ePDF:	978-1-3503-4198-2
	eBook:	978-1-3503-4199-9

Typeset by RefineCatch Limited, Bungay, Suffolk

To find out more about our authors and books visit www.bloomsbury.com and sign up for our newsletters.

Contents

List of Figures	vii
List of Tables	ix
Notes on Contributors	x
Acknowledgements	xiii
List of Abbreviations	xv

Introduction *Yasushi Hirai* — 1

Part One Memory and Mind

1. A Secret Connection Between the Dualities Actual/Virtual, and Intelligence/Intuition in Bergson's Metaphysics *Paul-Antoine Miquel* — 15
2. Bergson and the Rise of 'the Sciences of Memory' *Takeshi Miyake* — 25
3. Bergson's Dualism Today *Joël Dolbeault* — 43
4. Memory and History: Rereading Bergson from Ricœur *Masato Goda* — 53
5. 'A Long-accepted Foreigner, To Whom One Grants Refuge for a While': On Riquier's Interpretation of Bergson's Kantianism *Hisashi Fujita* — 63

Part Two Perception and Embodiment

6. Bergson, Gibson and the Image of the External World *Stephen E. Robbins* — 81
7. Bergson and Ecological Psychology: Memory of the Body and of the Universe *Tetsuya Kono* — 97
8. Defining Philosophy and Cognitive Psychology: Bergson and Bruner *Sébastien Miravète* — 107
9. What is the 'Thickness' of The Present? Bergson's Dual Perception System and the Ontology of Time *Yasushi Hirai* — 117
10. Affordance and Bergson *Tatsuya Higaki* — 133

Part Three Time and Duration

11. Neutral Monism, Temporal Experience and Time: Analytic Perspectives on Bergson *Barry Dainton* — 139
12. What Arranges Memories in a Line? *Takahiro Isashiki* — 161

13	Coexistence and the Flow of Time *Elie During*	175
14	Connection and Disconnection of Perception and Memory: Déjà vu, Bayesian and Inverse Bayesian Inference *Yukio-Pegio Gunji*	195
15	The Extensionalist View and Bergson's Notion of Contraction *Ryusuke Okajima*	213
16	We Bergsonians: The Kyoto Manifesto *Elie During and Paul-Antoine Miquel*	223

Index 243

Figures

0.1	PBJ 2015–2017 (*Matter and Memory*) Research Map. © Yasushi Hirai	7
2.1	An example of the card used in Smith's experiment (Smith, 1895)	26
2.2	Experimental psychologists in *Matter and Memory* © Takeshi Miyake	27
2.3	Schema of Lichitheim (Lichtheim 1885: 443)	29
2.4	Schema of Lichitheim (Lichtheim 1885: 478)	29
2.5	Schema of Charcot (Bernard 1885: 45)	30
2.6	Schema of Broadbent (Broadbent, 1879: 495)	31
2.7	Schema of Kussmaul (Kussmaul, 1884: 234)	31
2.8	Schema of Magnan (Skwortzoff, 1881, Planche I) captions added by the quoter	32
2.9	Schema of Moeli (Moeli, 1890: 380)	33
2.10	Schema of Wysman (Wysman, 1891: 37)	33
2.11	Schema of Freud (Freud, 1891: 83)	35
3.1	Libet's experiment, results. © Joël Dolbeault	48
6.1	Construction of a hologram. © Stephen E. Robbins	83
6.2	Reconstructing the image (i.e. the source of the original wave front). © Stephen E. Robbins	83
6.3	Modulating the reconstructive wave – from f_1 to f_2 © Stephen E. Robbins	84
6.4	The brain acts as a reconstructive wave, passing through the holographic field, specifying a source within the field. © Stephen E. Robbins	85
6.5	The 'ground'. © Stephen E. Robbins	86
6.6	Texture gradient. © Stephen E. Robbins	86
6.7	Flow field – a gradient of velocity vectors. © Stephen E. Robbins	87
6.8	Flow field – tau value for landing. © Stephen E. Robbins	87
6.9	The Gibsonian cube – a partitioned set of flow fields. © Stephen E. Robbins	88
6.10	Rotating ellipse – velocity vectors. © Stephen E. Robbins	88
6.11	Rotating wire cubes. © Stephen E. Robbins	89
6.12	Motion treated as a set of static points. © Stephen E. Robbins	91
6.13	The aging of the facial profile – a strain transformation applied to a cardioid. © Stephen E. Robbins	93
7.1	Bergson's inverted cone. Henri Bergson. Public domain	101
9.1	The diverging streams at the exit of the primary visual cortex. © Yasushi Hirai	117
9.2	The diverging streams at the superior colliculus. © Yasushi Hirai	117
9.3	Depiction A and B of contraction. © Yasushi Hirai	124
9.4	The triple structure of Bergsonian spacetime. © Yasushi Hirai	127
11.1	Cinematic model. © Barry Dainton	146

11.2 Retentional model. © Barry Dainton	147
11.3 Extensional model. © Barry Dainton	148
11.4 The main conceptions of contemporary analytic philosophy of time. © Barry Dainton	153
12.1 Bergson's inverted cone. Henri Bergson. Public domain	163
12.2 The order of arrival. © Takahiro Isashiki	167
12.3 The order of events in linear time. © Takahiro Isashiki	168
14.1 Perception of P established in the location of P from the virtual first-person perspective. Emotion and/or *qualia* is formed in the inside. © Yukio-Pegio Gunji	197
14.2 An exchange of the past makes of two past histories coexistence, which could create a feeling of déjà vu. © Yukio-Pegio Gunji	202
14.3 Inverse cone generated by the synthesis of multiple triangles and planes. © Yukio- Pegio Gunji	203
14.4 Left: inverse cone implemented by Bayesian and inverse Bayesian inference. Right top: the probability of taking a red ball from the optimum hypothesis estimated only by Bayesian inference potted against the time step. The probability of taking a red ball is changed from 0.7 to 0.2 when plotted against time. Right bottom: the probability of taking a red ball of the optimum hypothesis estimated by Bayesian and inverse Bayesian inference when plotted against time. © Yukio- Pegio Gunji	207
15.1 The succession of material moments © Ryusuke Okajima	217
15.2 The contraction in ECM1 © Ryusuke Okajima	217
15.3 The contraction in ECM2 © Ryusuke Okajima	218

Tables

5.1 The divisions of the *Critique of Pure Reason*. © Hisashi Fujita 66
5.2 Correspondence between Bergson's works and *Critique of Pure Reason* (Riquier). © Hisashi Fujita 73

Notes on Contributors

Barry Dainton is Professor of Philosophy at the University of Liverpool. He specializes in metaphysics and philosophy of mind, and his also interested in the philosophical advances in science and technology. He is one of the co-directors of the Olaf Stapledon Centre for Speculative Futures. His previous publications include *Stream of Consciousness* (2000), *The Phenomenal Self* (2008), *Time and Space* (2010) and *Self* (2014). He is co-editor of *The Bloomsbury Companion to Analytic Philosophy* (Bloomsbury 2014) and *Minding the Future: Artificial Intelligence, Philosophical Visions and Science Fiction* (2021).

Joël Dolbeault is an independent researcher. He studied the history of philosophy and the philosophy of science at the University of Paris-Sorbonne, and obtained a PhD at the University of Lille, with Frédéric Worms as his thesis director. His research aims at confronting Bergson's main metaphysical ideas with contemporary ideas in philosophy and science. He is particularly interested in Bergson's panpsychism, in its relation to contemporary panpsychism. He has published several articles on this subject in French and in English.

Elie During is Associate Professor in philosophy at University Paris Nanterre. His publications include several critical editions: Henri Bergson's *Duration and Simultaneity* (*Durée et simultanéité*, 2009), Paul Langevin's 1911 lectures on relativity theory (*Le Paradoxe des jumeaux: deux conférences sur la relativité*, 2016), Gaston Bachelard's *Dialectic of Duration* (*La Dialectique de la durée*, 2022). Among his recent contributions on Bergson is 'Time as Form: Lessons from the Bergson-Einstein Dispute', published in *Einstein vs. Bergson: An Enduring Quarrel on Time* (A. Campo and S. Gozzano eds, 2021).

Hisashi Fujita is Professor of Philosophy at Kyushu Sangyo University, Japan. He is author of *Bergson: An Untimely Philosophy* (2022), co-translator in Japanese of Bergson's *Histoire de l'idée de temps* (2019), co-editor of *Considérations inactuelles: Bergson et la philosophie française du XIXe siècle* (2017) and of *Disséminations de L'évolution créatrice de Bergson* (2012). Among his publications are 'Diremption and Intersection: The Violence of Language in Bergson and Sorel', *Parrhesia. A Journal of Critical Philosophy* (2022), 'Télépathie: la recherche psychique de Bergson et la métapsychologie de Freud' in *Bergson et Freud* (2014).

Masato Goda is Professor at Meiji University, Tokyo, Japan. His main research interests lie in modern history of philosophical ideas in Europe as well as of Jewish thoughts. His major publications include *Levinas* (2000) and *Jankélévitch. Rhapsody of Boundaries* (2004), among others. He is also the translator of Bergson's major works.

Yukio-Pegio Gunji is a professor in Department of Intermedia Art and Science, Waseda University. His current research interests are creativity in human and animals, psychological origin of quantum logic, origin of consciousness, bio-inspired computing. His recent books are *Natural Born Intelligence* (2019), *Something Arriving* (2020).

Tatsuya Higaki is Professor at the University of Osaka, Faculty of Human Sciences. His specialty is contemporary French philosophy, Japanese philosophy and twenty-first century development of French philosophy. He is the author of many books, on Deleuze, Foucault, Bergson, Kitaro Nishida among others. His book *Nishida Kitaro's Philosophy of Life* is translated into English (2020).

Yasushi Hirai is Professor in the Faculty of Letters at Keio University, Japan. He specialises in philosophy of time, mind and memory, with a focus on Bergson and Leibniz. He is chief director of PBJ (Project Bergson in Japan) and a board member of Société des Amis de Bergson and Global Bergsonism Research Project. He translated Bergson's *Time and Free Will*, *Histoire de l'idée de temps* and *Histoire des theories de la mémoire* into Japanese. He is co-editor of *The Anatomy of Bergson's* Matter and Memory (2016), *Diagnoses of Bergson's* Matter and Memory (2017) and *Rebooting Bergson's* Matter and Memory (2018).

Takahiro Isashiki is a specially-appointed professor in the Faculty of Economics at Nihon University. Born in 1956, he majors in philosophy and took his doctorate in Wittgenstein Studies at the University of Tokyo. His main publications include *What Happens to Us after We Die?* (2019), *Metaphysics of Temporal Modality* (2010).

Tetsuya Kono is Professor at Rikkyo University, Tokyo, Japan. His main research interests lie in phenomenology, philosophy of mind, and philosophical psychology. He is also involved in the field of environmental philosophy and environmental ethics. His major publications include *Ma'ai: An Inquiry of Ecological Phenomenology* (2022), *Everything Will Revert to Wild Someday: Phenomenology of Environment* (2016), *Phenomenology of Body and Special Needs Education* (2015) and *Phenomenology of Boundaries* (2014).

Paul-Antoine Miquel is Professor of Contemporary Philosophy at the Jean Jaurès University in Toulouse. His research focuses on French philosophy and the philosophy of biology. He is the author of several works on Bergson, Simondon, Canguilhem, and he is the editor of the works of Bergson published by Garnier Flammarion in France.

Sébastien Miravète has a PhD in philosophy and a PhD in cognitive psychology. He is an associate researcher at the philosophy laboratory ERRAPHIS of the University of Toulouse 2 and an associate researcher at the psychology laboratory EPSYLON of the University of Montpellier 3. He works on the concept of cognitive structure from an ontological (Bergson), ecological (linking different scales of reality) and cognitivist (cognitive load theory) perspective.

Takeshi Miyake is a professor in the faculty of educations at the University of Kagawa. He pursues research into the formation of Bergson's philosophical concepts and their relations with the development of sciences in the same period. He has published his doctoral thesis: *Interactions Between the Philosophy of Bergson and Positive Sciences* (2012).

Ryusuke Okajima is an Associate Professor in French Philosophy at Niigata University. He received his PhD degree in philosophy from Keio University, Japan, in 2020. His current research interests are Bergson's theory of time and space, the history of nineteenth-century French philosophy of science, and contemporary theories of temporal consciousness. His recent paper is 'On the Problem of Quality and Quantity in Early Bergson from *Time and Free Will* to *Matter and Memory*' (2020).

Stephen E. Robbins is a former software company executive, now retired. He did his PhD work on Bergson and J.J. Gibson at the University of Minnesota under Robert E. Shaw, founder of the *Journal of Ecological Psychology*, and has published roughly twenty academic journal articles around this subject. The latest being 'LSD and Perception: The Bergson-Gibson model of direct perception and its biochemical framework', *Psychology of Consciousness*.

Acknowledgements

First and foremost, I would like to thank all the philosophers who agreed to participate in this project and who – together with, and sometimes 'against', Bergson – demonstrated the ideal way to wrestle with the matter itself. Many of them were willing to respond to my detailed preliminary requests regarding the content of the discussions, even though I had never met some of them before. The three days of the symposium were replete with heated debates, occupying over half of the total program time. I cannot express the extent to which I was encouraged by the words that the participants sent me upon their return. I will remain ever proud to have been able to convene this line-up.

This international collaborative research is part of a long-running project ongoing since 2007 under the name of PBJ (Project Bergson in Japan). The current members of PBJ are Shin Abiko, Hisashi Fujita, Masato Goda, Tatsuya Higaki, Kazuyori Kondo, Takeshi Miyake, Tatsuya Murayama, Yasuhiko Sugimura and Hirai Yasushi. This mission would not have been possible without the active support of these leading philosophers in Japan.

In particular, I would like to express my special thanks to Professor Shin Abiko and Hisashi Fujita, who were central to the launch of the project. When I took over the project in 2015, the international trust that is essential for this kind of collaboration had already been built. Without this, I would not have been able to fulfil such an ambitious initiative. I also must mention Professor Osamu Kanamori, who passed away in the middle of the project. I would also like to thank Naoki Sugiyama and Tatsuya Murayama, two of Japan's leading Bergson scholars, for their encouragement and useful advice at various times.

The various communities of philosophical discussion, including the Bergsonian Philosophy Study Group, have led me here. The guidance of my advisor, Professor Atsushi Fukui, continues to be the driving force behind my research. It was the late Professor Toshio Ishii who discovered me as a Bergsonian. I would like to express my gratitude to him.

The original event would not have been possible without the excellent interpreters Takafumi Ishiwatari, Katsunori Miyahara, Masumi Nagasaka, and Jun Otsuka, who are all now at the forefront of their respective fields. The young researchers who prepared high-quality translations and the part-time workers at the venue, made it possible to have such fruitful discussions. Thank you very much.

In grateful acknowledgement, this study is supported by the Grant-in-Aid for Scientific Research (B) under Grant No. 15H03154 and 19H01190 from the Japan Society for the Promotion of Science and the Fukuoka University Research Fund No. 197002 and 203004. I would also like to extend my special thanks to the staff of Fukuoka University who were involved in the tedious administrative work. Mr Hiroshi Seito of the Japanese publishing house Shoshi Shinsui, not only accepted the publication of the original version in Japanese when our challenge was still at its early stages at the time, but also readily agreed to publish the English version of the work. We are all grateful to him.

Finally, let me express my warmest thanks to my parents, family, friends, colleagues and students. It is thanks to you all that I have been able to lead a fulfilling research life.

Abbreviations

In the text, references to Bergson's work are indicated by the following abbreviations and English translations, with the corresponding French original page numbers provided in brackets, where available.

CE   Bergson, Henri (1911), *Creative Evolution*, trans. A. Mitchell, London: Macmillan. *L'Évolution créatrice* (1907), édition critique réalisée par Arnaud François (2007).

CM   Bergson, Henri (1946), *The Creative Mind*, trans. M.L. Andison, New York: Philosophical Library. *La Pensée et le Mouvant* (1934), édition critique réalisée par Arnaud Bouaniche, Arnaud François, Frédéric Fruteau de Laclos, Stéphane Madelrieux, Claire Marin, Ghislain Waterlot (2009).

DS   Bergson, Henri (1965), *Duration and Simultaneity*, trans L. Jacobson, Indianapolis: Bobbs-Merrill. *Durée et simultanéité* (1922), édition critique réalisée par Élie During (2009).

L   Bergson, Henri (1914), *Laughter: Essay on the Meaning of the Comic*, trans. C. Brereton and F. Rothwell, New York: MacMillan. *Le Rire* (1900), édition critique réalisée par Guillaume Sibertin-Blanc (2007).

ME   Bergson, Henri (1920, *Mind-Energy*, trans. H. Wildon Carr, New York: Henry Holt and Co. *L'Énergie spirituelle* (1919), édition critique réalisée par Arnaud François, Élie During, Stéphane Madelrieux, Camille Riquier, Guillaume Sibertin-Blanc, Ghislain Waterlot (2009).

MM   Bergson, Henri (1911), *Matter and Memory*, trans. N.M. Paul and W.S. Palmer. London: George Allen and Unwin. *Matière et mémoire* (1896), édition critique réalisée par Camille Riquier (2008).

TFW   Bergson, Henri (1910), *Time and Free Will*, trans. F.L. Pogson, London: George Allen & Unwin. *Essai sur les données immédiates de la conscience* (1889), édition critique réalisée par Arnaud Bouaniche (2007).

TS   Bergson, Henri (1935), *The Two Sources of Morality and Religion*, trans. R. Ashley Audra and C. Brereton, New York: Henry Holt, 1935. *Les Deux Sources de la morale et de la religion* (1932), édition critique réalisée par Frédéric Keck et Ghislain Waterlot (2008).

Introduction

Yasushi Hirai
Keio University

This collection of essays is based on the international symposium titled 'The Anatomy of *Matter and Memory*: Bergson and Contemporary Theories of Perception, Time, and Mind' held in Tokyo and Kyoto, Japan, in December 2015. It includes essays based on the manuscripts presented by all twelve presenters, three short chapters inspired by them (chapters 5, 10 and 15) and the 'manifesto' (see below, chapter 16).

As the title of the symposium indicates, this project aims to thoroughly examine the philosophical potential of Henri Bergson's masterpiece *Matter and Memory* – said to be his most difficult book – by examining it from the perspective of contemporary analytical metaphysics and scientific theory. In fact, given the science-informed metaphysical[1] nature of the original work, this is not the first attempt to establish such a connection between Bergson and modern science. Our study would not have been feasible without those previous studies conducted in the 1980s and 1990s (Gallois and Forzy 1997, Mullarkey 1999, Papanicolaou and Gunter 1987). However, because of the changes in the field described below, the present attempt introduces new meaning that has remained underexamined in previous studies.

The challenge of *Matter and Memory*

Before explaining these circumstances, I will first examine the complex structure of the original book. Simply put, *Matter and Memory* presents an attempt to completely reconstruct the *mind-body problem* 'in terms of time (MM, 77[74]; 294[248])'. Therefore, it might prove insufficient to only examine the main subjects that Bergson attempts to undertake. The book is composed of a complex combination of 'dense' themes that are troublesome on their own, such as the 'mind-body problem', 'theory of knowledge' and 'theory of time'. Naturally, the book includes many philosophical references, such as to Descartes, Berkeley, Leibniz and Kant.

However, a strong characteristic of this book is that it draws on these apparently metaphysical points of view as an extension of meticulous empirical research on the following: the biology of neural tissue; the pathology of apraxia and aphasia; the psychology of attention, memory and general ideas; and the physics of forces and atoms. As many as seventy scientists[2] are mentioned within only the second chapter of the book where he develops his theory of recognition. How can such a detailed empirical scientific inquiry and bold metaphysical thesis be brought together in one book? Bergson's reasoning is not immediately obvious. As Higaki urges in his chapter, this question remains unexplored. We can at least be sure that the relationship between the two is neither a simple induction from various scientific fields nor a top-down application of arm-chair philosophical notions;[3] in fact, Bergson himself consciously criticized such methodology.[4] Moreover, many aspects of methodological opaqueness have been barriers to its acceptance: basic terms (e.g. 'perception', 'representation', 'image', 'memory', 'existence' and 'past') are so vaguely tangled that their referents and connotations cannot be easily distilled. Each usage adopted by Bergson seems to be bizarre and perplexing,[5] and the underlying logic that drives the argument continues to escape our grasp.[6]

In fact, if we compare the status of Bergsonian and phenomenological research in terms of how well the theory is understood by non-specialists, the difference is obvious. Although it was formed in the same period as Bergson and was also originally labelled a 'continental' philosophy, phenomenology has already entered the dimension of collaborative inquiry with many scientific fields (Gallagher and Schmicking 2010, Gallagher and Zahavi 2008, Zahavi 2003, 2012), largely because of the efforts of Gallagher, Zahavi and others in recent years. It has made numerous contributions to various fields. Thus, we might[7] say that *Matter and Memory* has not yet even reached the realm of academic dissemination.

High-dimensional science metaphysical puzzles

We believe that one of the reasons for this lies in the unusual complexity of *Matter and Memory*. The following is an attempt to arrange the individual (and seemingly bizarre) claims contained in *Matter and Memory* in list form:

1. Theses on matter and consciousness, such as 'perception is a part of matter', 'matter is a neutralized consciousness', 'the brain cannot in principle produce either *qualia* or representations', '*qualia* occupy spatial extension' and 'delayed neural responses trigger consciousness'.
2. Theories on cognition, such as 'bodily movement is the basis of recognition', 'recall does not imply recognition', 'in cognition, details are accomplished after the whole' and 'interpretation cannot be established solely bottom-up'.
3. Claims about the past and memory, such as 'dreams are the relaxation of memory', 'a perception does not become a memory', 'the mind is made of the past', 'the same memory can be recalled in countless different ways' and 'memory is not stored in the brain'.
4. Issues of time and freedom, such as 'the present is the latest cross-section of the real', 'all the past remains', 'motion is indivisible' and 'a timescale-shift brings freedom'.

This summation of Bergsonian theses implies that they surpass the control of the conventional paradigm. Even a Bergson scholar would be almost dizzy to see them listed in this manner. Moreover, these multiple lines of thought are intricately intertwined, forming tight knots everywhere. Even though over 120 years have passed since the publication of the book in 1896, there is an unusual gap between the lack of knowledge of these particular issues (even compared to Bergson's other works) and the nominal fame of the Nobel Prize winner's main work. This fact cannot be attributed simply to unfortunate historical circumstances.[8] It is necessary to recognize at the outset that it clearly also stems from the extreme complexity of its content.

However, the situation is changing. Thanks to the relevant fields of science and analytical metaphysics, which have made drastic progress in the first two decades of the twenty-first century, it is finally becoming feasible to treat the theses of *Matter and Memory*, which have long remained difficult to digest, as truly verifiable. Moreover, the key to unravelling these complications is, once again, 'time'. In both science and philosophy, there were two stages of development: first, the relevant fields themselves emerged, and then, the importance of the question of 'time' came to be recognized as well. I trace these two distinct domains sequentially below.

Consciousness and time in the sciences since the 1980s

There has been dramatic progress especially in the so-called 'science of consciousness'. The pioneering work of Francis Crick, Christof Koch, Gerald Edelman and others in the early 1990s opened the door to the scientific study of consciousness, which had long been considered taboo. With the help of subsequent breakthroughs in brain imaging equipment, the field has undergone rapid development. In this context, scientists such as Michael S. Gazzaniga, Giulio Tononi and Stanislas Dehaene made a series of attractive hypotheses and theories in the 2010s, approaching the generative condition of consciousness. Simultaneously, the engineering view of the physical basis of intelligence has been qualitatively transformed because of the emergence of artificial intelligence with deep learning since 2012.[9] In particular, the advent of the 'constructivist approach',[10] which was hitherto technically impossible, has yielded several perspectives that could not have been anticipated at a time when consciousness could only be the object of phenomenological description and experimental analysis.

What has suddenly emerged as a challenge is the re-examination of the concept of time. Cognitive neuroscientist Gazzaniga pointed out in his 2011 book that the concept of 'timescale' plays an important role in elucidating the mechanisms of consciousness (Gazzaniga 2011: 141, Koch 2011: chap. 7).[11] Tononi and Massimini, proponents of the integrated information theory, found that temporal delays in response, rather than the synchronous firing of nerves in the brain, conditioned consciousness.[12] Daniel Kahneman, the founder of behavioural economics, proposed a dual-process model of thought in his best-selling book *Fast and Slow* (also in 2011), arguing that sequential, 'slow' processes are more conscious than parallel, rapid ones.[13] Physicist Michio Kaku regards the various levels of consciousness, from matter to man, as a gradual extension of a feedback loop and states that a significant extension to the time dimension has made advanced consciousness possible (Kaku 2014). These latest claims, which have all been made in unison, cannot help but throw light back on the central thesis of *Matter and Memory*, which is that consciousness is triggered by slowness or temporal prolongation, and that matter is unconscious because its timescale is so short that it can be approximated by simultaneity.[14] Lee Smolin, a proponent of the loop quantum gravity theory, also argues in *Time Reborn*, published in 2013, that it is important to restore 'flowing time' to physics, which has long been dominated by the block universe theory (Smolin 2013).[15] In the field of cognitive psychology, through the issues raised by Ned Block and others concerning phenomenal consciousness and access consciousness (Block 2011, Nishimura 2014, Sato 2014), it is becoming generally recognized that the elucidation of the structure of the temporal experience largely affects the way we conceive consciousness.[16] In this way, through the many attempts at the scientific clarification of consciousness, the key issues unexpectedly converge on the issue of time – an issue that bears much in common with Bergson's *Matter and Memory*.[17]

Mind and time in analytic metaphysics

Another important element to this background is the rise of 'analytic metaphysics' and the situation regarding philosophy of time. In the mid-1990s, sometime after the emergence of the 'science of consciousness', the 'philosophy of mind' as a genre emerged from within the field of analytic philosophy which had had a strong materialistic tendency hitherto, over the hard problem of *qualia* and mental representations. The mind-body problem was thus resurrected as one of the current issues in philosophy. It is not the mind alone that has gained attention.[18] The analytical approach, based on thorough conceptual refinement, has proven itself effective in dealing with such 'metaphysical' issues as being, God, freedom, universals, modality and beauty, among others, which the earlier 'analytic philosophy' would have criticized. This has led to a comprehensive reinterpretation of early modern philosophy and phenomenology with sophisticated frameworks for argumentation and high-resolution concepts. It signifies a departure from the sterile 'Anglo-American versus Continental' rivalry of the past.

Unfortunately, however, with regard to the concept of time, the separation of the two traditions was prolonged. The theory of time in the Anglo-Saxon world, which originated in John McTaggart's paper 'The Unreality of Time' (1908) and was formed independent of Bergson and phenomenology, showed early results through several controversies. By the turn of the twentieth century, it had already been received by common physicists. Therefore, it was ready to provide various conceptual tools (endurance vs. perdurance, presentism, eternalism, growth block universe, A-theory and B-theory, among others) necessary to treat time in a logically clear manner. However, the twentieth century passed without any real dialogue between the analytical and the continental theories of time.[19]

This changed with the publication of Barry Dainton's *Stream of Consciousness* in 2000. In addition to the conventional ontological controversy about the structure of 'time' in itself, Dainton's book opened up a new realm of debate about the structure of 'temporal experience' where falls the central notion of Bergson's philosophy, *durée* (duration in French). In this context, views are divided between extensionalism and retentionalism. Simply stated, the latter claims that the temporal extension of consciousness is extended only subjectively whereas the former affirms that it is extended in the real temporal dimension. Dainton's clear stage setting has caused a heated debate among temporal theorists since around 2011.[20] Moreover, it is noteworthy that Dainton drew this line of thought through the careful re-reading of traditional texts of William James and Edmund Husserl, both of whom had a direct and close ideological influence on Bergson. Thus, the stage is now set for a debate in which the English and French temporal traditions – which emerged at roughly the same time across the Strait of Dover but have strangely been at odds with each other – can start a real conversation without being distracted by superficial differences in vocabulary or unfortunate historical biases.[21]

The science of metaphysics

I would add that these radical changes in both science and philosophy are not, of course, unrelated to each other. The scientific situation of the mind is definitely shifting to the next paradigm, prompting a serious reformulation of epistemology and the mind-body problem in philosophy. In response to this situation, *Scientific Metaphysics*,[22] edited by Ross, Ladyman and Kincaid (2013) was published. Some commentators have explained this movement by comparing it with 'metametaphysics', which shows a tendency towards speculation (Healey 2013). As noted at the beginning, Bergson was clearly oriented towards a science-based metaphysics.[23] By contrast, however, Bergson did not intend to simply 'naturalize' metaphysics in the sense of current conventional natural science. How can the science and philosophy of consciousness advance to a new frontier without questioning the very meaning of 'nature' in the first place? What is required is not the reduction of philosophical problems to science, nor is it for philosophy to unilaterally make sense of science. *Matter and Memory* challenges us to rethink how time plays a decisive role in nature, thereby encouraging us to elaborate a novel framework for both consciousness and nature in terms of *durée* and *image*. Now that we have the tools to 'anatomize'[24] the scientific metaphysical complication of *Matter and Memory*, we need to examine the individual theses listed above in depth.

It is probably no coincidence that several commentators began to notice the contemporary value of Bergson's concepts around the same time. Riggio wrote the following in 2016:

> Throughout *Matter and Memory*, Bergson conducts himself in a manner that we today would consider a model of the scientifically – informed philosopher: a builder of innovative arguments and concepts that break decisively from tired stereotypes and out of date common sense about reality, a philosophical articulation of a detailed, nuanced, and attentive understanding of contemporary scientific discoveries.
>
> Riggio 2016: 216

David Kreps, who published *Bergson, Complexity, and Creative Emergence* in 2015, points out, with a wealth of concrete examples, that recent findings reveal that the criticisms of Bergson based on the sciences of the twentieth century have not always been accurate. Against this background, he states, the foresight of Bergson's theses (which had not been recognized before) has come to stand out, so to speak, in a roundabout way. Kreps describes the situation:

> Bergson's ideas are enjoying something of a revival in various circles. He has, in the past, had many critics; and there remain – in this author's eyes, at least – some elements of his work that have not stood the test of time... Nonetheless, there have been developments in both philosophy – in particular, poststructuralism – and in scientific endeavour – for example, some of the discoveries of contemporary neuroscience – where Bergson's ideas have, despite early criticism, proved far more accurate than his detractors'.
>
> Kreps 2015: 1–2

The same point that the earlier criticism of Bergson's theses has been limited has been made by Inaga and Barnard.[25] These remarks univocally indicate that the current movement of re-evaluation is qualitatively different from that which was once sparked within philosophy due to Deleuze's powerful interpretation.

Dainton, who has also written a textbook on modern metaphysics, has hitherto not made any comprehensive reference to Bergson (at that moment). In the beginning of his essay in this book, however, he makes the following in-depth assessment of Bergson:

> To say that Bergson's work has been largely overlooked in recent analytical philosophy would be an understatement. I will not hazard a guess as to whether this will change in the years to come, but one thing is clear: this situation ought to change. Some of the issues close to the centre of current analytical concerns – issues concerning the relationship between the mental and the physical, the nature of time, the ways in which consciousness is temporal – Bergson not only had interesting things to say, some of what he had to say may well turn out to be true.[26]

Towards a theoretical application of Bergsonian philosophy

It should also be pointed out that if all this 'foresight' is not the effect of confirmation bias due to 'blind faith', then Bergson's philosophy will be not only an object of interest in the history of thought, but also an active theoretical component that can present effective and attractive alternatives in contemporary debates. The authors present in this collection are unanimous in this attitude, as Dainton's testimony illustrates:

> I think something else is clear as well: Bergson solution to the problem of consciousness is not solely of historical interest. For anyone interested in a unified, monistic, world-view, one where the mental and the physical are not fundamentally

different in kind, then the approach pioneered by Bergson ... is well worth considering.

p. 144

This book is divided into three parts, each of which contains essays in the order in which they were presented at the symposium held over three days. Finally, this book includes a manifesto written by Paul-Antoine Miquel and Elie During, 'We Bergsonians: The Kyoto Manifesto'.

This manifesto, composed during their stay in Japan and after their return, is a powerful statement of the direction of this project and a proposal for the future research. I would like to take this opportunity to express my gratitude to both authors and the editor of the journal *Parrhesia* for their permission to reproduce it here in this volume. I hope that many people, not only Bergson scholars, will read this book and pause to consider the relationship between science and metaphysics as well as that between classical texts and contemporary issues. We hope that our attempts will spark people's insights and lead to various dialogues.

Our project to revisit *Matter and Memory* did not end with this book. Symposia have been held for three consecutive years, including this one, all of which have been published in Japanese, bringing the total number of essays to fifty.

It has continued to open up connections to new fields: with analytic philosophy (temporal ontology, panpsychism/panqualityism, morality and formal ontology);[27]

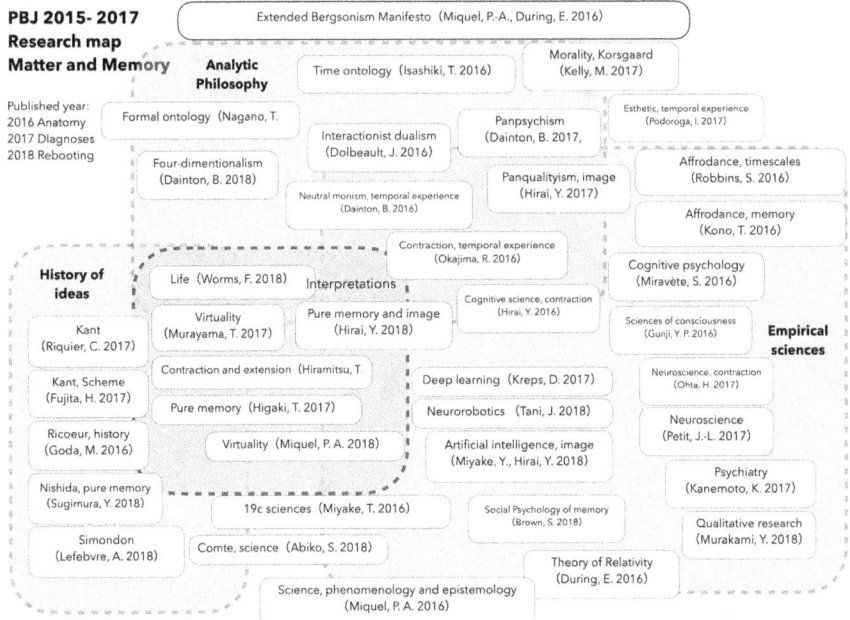

Figure 0.1 PBJ 2015–17 (*Matter and Memory*) Research Map. © Yasushi Hirai.

with science (neuroscience, psychiatry, social psychology, artificial intelligence and neuro-robotics);[28] and with the history of philosophy (Kant, Simondon, Kitaro Nishida and Comte).

In recent years, similar attempts have become increasingly widespread,[29] and it should be said that the period of innovation in Bergsonian studies continues unabated.

Notes

1. He uses the term '*métaphysique positive*' to refer to his philosophical methodology, where the adjective 'positive' is in the sense of 'positive science'. See Bergson's article 'Psychosomatic Parallelism and Empirical Metaphysics' (Bergson 1972: 463–504).
2. See Miyake's chapter in this book.
3. See the manifesto by Miquel and During. For example, as many scholars have unanimously acknowledged, the theory of *durée* in Bergson's first book, *Time and Free Will*, makes an important ontological extension (a commitment to the reality of a 'past' beyond the phenomenal present) through its consideration in *Matter and Memory*.
4. See Bergson's articles 'Introduction I and II', 'Philosophical Intuition', and others in *Creative Mind*.
5. See Miravète's chapter in this book.
6. See Hirai's chapter in this book.
7. For a systematic repudiation of the true dialogue between phenomenology and Bergson in the past, and the possible contribution of Bergson's philosophy to phenomenology, see Kelly (2010).
8. A decade after the publication of *Matter and Memory*, Bergson's fame, which made him the darling of the age with his *Creative Evolution*, ironically promoted a superficial dissemination of his philosophical works. The French word '*élan vital*' (introduced as a foreign word even in English-speaking countries) evokes a kind of spell-like suspension of thought, and it was not unusual for it to be confused with 'finalism', which he himself systematically criticized by name. In the history of ideas, we should not forget that the 'overcoming' by Merleau-Ponty also had a great influence (see Kono's chapter in this book). After that, Deleuze's *Le bergsonime*, published in 1968, caused a re-evaluation in the philosophical circles as a 'philosophy of difference', but it did not lead to a movement to develop the scientific metaphysical scope of *Matter and Memory*. It cannot be said that the Sokal affair was not involved.
9. Deep learning made its shocking debut when it won the ILSVRC, a global image recognition competition, by more than ten per centage points over traditional machine learning.
10. The physical and ethical limits of dealing with a living human being are not present in the simulation. It is different from simply physiological imitation (omnipotent architecture) and is an attempt to find an equivalent as a system. See Nakajima (2016), Matsuo (2016) and Taniguchi (2014).
11. See also Dolbeault's chapter in this book.
12. The theory of integrated information states that it is essential for the development of consciousness that the amount of information and its integration, which are usually in a trade-off relationship, come together at just the right time. Massimini and Tononi showed that this condition is achieved precisely by temporal prolongation, that is, 'sparse response delays rather than simultaneous firing of nerves'. See Massimini and

Tononi (2013, 143–7) and Tononi (2004). Koch (2011, 129) also refers to the 'timescale' in Tononi's theory.
13   On the trend of the cognitive science of thought, see Ohta and Oguchi (2014). In addition, for the 'zombie system' of the 'fast' side, see Koch (2011).
14   The prototype of the idea of matter as 'spirit without memory' is often attributed to Leibniz ('Abstract Kinetics', 1670 Gerhardt's edition, volume 4, 230), but it is important to note that in Bergson it is not just an 'idea', but is implemented in a concrete, empirical way.
15   Some physicists were influenced by Bergson early on, such as Watanabe (1951) and Prigogine (1980). The Japanese edition of Prigogine's *Being and Becoming* published in 1984 has an additional chapter 10, which does not exist in the English edition, where he attempts to discuss 'uncompressible duration' of Bergson and Whitehead in physical terms. On physics and Bergson's theory of time, see also Čapek (1971) and Gunter (1969, 1991).
16   Sato (2014) offers a lucid and informative overview of this controversy.
17   Of course, it is also essential to crosscheck with C.S. Peirce, who shares the same will of 'scientific metaphysics', and Dewey, whose key concept is 'delayed reaction'. See Nubiola (2014) for the complicated situation in which the famous expression 'scientific metaphysics' that was used on the spine of the sixth volume of the collected works of C.S. Pierce was used only once in his known writings (Peirce 1931–58, volume 8, 284).
18   For more details, please refer to Barry Dainton's chapter in this book.
19   With only a few exceptions. See Williams (1998).
20   Although the first edition of *Stream of Consciousness* was published in 2000, the controversy under the banner of extensionism has only intensified since Dainton (2011, 2012, 2013, 2014).
21   In addition, Hirai (2011) experimentally attempted the verification of Bergson's time ontology by the analytic system.
22   See also note 17 for Peirce's version of it.
23   See note 1 above. See also the manifesto by Miquel and During in this book for a detailed discussion of this point.
24   The original event was entitled 'The Anatomy of Bergson's *Matter and Memory*'.
25   The following paper by Masahiro Inaga is in line with the idea of this symposium and includes a detailed examination of its relationship with previous studies, which I have to omit here. Please refer to Barnard (2011) and Inaga (2013).
26   Referring to Chalmers' panpsychism, he notes, 'It is clear that the proponents of panpsychism are not far from Bergson's, for in *Matter and Memory* Bergson argues that it is essential to the avoidance of dualism to recognize the existence of sensory qualities in the external world'.
27   We held the cross-disciplinary event 'Rebooting Panpsychism' in Tokyo in December 2019, involving Russell scholars, Whitehead scholars and analytic philosophers, where the venue was almost full; it bore fruit in the June 2020 issue of *Gendai Shiso*, Japan's leading journal of thought, as a special issue on panpsychism. See Takamura and Suzuki (2020), Iimori (2020) and Hirai (2020).
28   For dialogues with the field of artificial intelligence using an extended Bergsonian framework, see Miyake and Hirai (2018); Tani, Miyake and Hirai (2018); Taniguchi, Hirai, Tsuda and Miyake (2021a, 2021b).
29   See for example Bunnag (2019), Dainton (2022), Deppe (2016, 2022), Fischer (2021), Moravec (2019, 2022), Mutch (2016), Nyíri (2014), Olma (2007), Riggio (2016), Sinclair (2019) and Wolf (2021).

It is gratefully acknowledged that this study is partially supported by the Grant-in-Aid for Scientific Research (B) under Grant No. 15H03154 and 19H01190 from the Japan Society for the Promotion of Science and the Fukuoka University Research Fund No. 197002 and 203004. I would like to thank Editage (www.editage.com) for English language editing.

References

Barnard, W. (2011), *Living Consciousness: The Metaphysical Vision of Henri Bergson*, Albany: State University of New York Press.
Bergson, H. (1972), *Mélanges*, Paris: Presse Universitaire de France.
Block N. (2011), 'Perceptual Consciousness Overflows Cognitive Access', *Trends in Cognitive Sciences*, 15 (12): 567–75. https://doi.org/10.1016/j.tics.2011.11.001
Bunnag, A. (2019), 'The Concept of Time in Philosophy: A Comparative Study between Theravada Buddhist and Henri Bergson's Concept of Time from Thai Philosophers' Perspectives', *Kasetsart Journal of Social Sciences*, 40 (1): 179–85, https://so04.tci-thaijo.org/index.php/kjss/article/view/239838
Čapek, M. (1971), *Bergson and Modern Physics: A Reinterpretation and Re-evaluation*, Dordrecht-Holland: D. Reidel Publishing Company.
Dainton, B. (2000), *Stream of Consciousness: Unity and Continuity in Conscious Experience*, London: Routledge.
Dainton, B. (2011), 'Time, Passage and Immediate Experience', in C. Callender (ed.), *Oxford Handbook of Philosophy of Time*, 382–419, Oxford: Oxford University Press.
Dainton, B. (2012), 'Time and Temporal Experience', in A. Bardon (ed.), *The Future of the Philosophy of Time*, 123–48, Routledge.
Dainton, B. (2013), 'The Phenomenal Continuum', in D. Lloyd and V. Arstila (eds), *Subjective Time: The Philosophy, Psychology, and Neuroscience of Temporality*, 101–37, Cambridge, MA: MIT Press.
Dainton, B. (2014), 'Flow, Repetition and Symmetry', in N. Oaklander (ed.), *Debates in the Philosophy of Time*, 175–212, London: Routledge.
Dainton, B. (2022), 'Irreducibility, Indivisibility, and Interpenetration', in Y. Wolf and M. Sinclair (eds), *The Bergsonian Mind*, 393–416, New York and Abingdon: Routledge.
Deppe, S. (2016), 'The Mind-Dependence of the Relational Structure of Time (or: What Henri Bergson Would Say to B-Theorists)', *Kriterion – Journal of Philosophy*, 30 (2): 107–24.
Deppe, S. (2021), 'Combining Tense and Temporal Extension: The Potential of Bergson's "Qualitative Multiplicity" for Conquering Problems of (Analytic) Time Metaphysics', *Bergsoniana*, 1. https://journals.openedition.org/bergsoniana/298
Fischer, F. (2021), 'Bergsonian Answers to Contemporary Persistence Questions', *Bergsoniana*, 1, http://journals.openedition.org/bergsoniana/448
Gallagher, S. and D. Schmicking, eds (2010), *Handbook of Phenomenology and Cognitive Science*, New York: Springer.
Gallagher, S. and D. Zahavi, eds (2008), *The Phenomenological Mind,* Abingdon: Routledge.
Gallois, P. and G. Forzy, eds (1997), *Bergson et les Neurosciences*, Paris: Institut Synthélabo.
Gazzaniga, M.S. (2011), *Who's in Charge?: Free Will and the Science of the Brain*, New York: HarperCollins.
Gunter, P.A.Y. (1969), *Bergson and the Evolution of Physics*, Knoxville: University of Tennessee Press.

Gunter, P.A.Y. (1991), 'Bergson and Non-Linear Non-Equilibrium Thermodynamics: An Application of Method', *Revue Internationale de Philosophie*, 45 (177 (2)): 108–21.
Healey, R. (2013), 'Review of *Scientific Metaphysics*', *Notre Dame Philosophical Reviews*, http://ndpr.nd.edu/news/41185-scientific-metaphysics/
Hirai, Y. (2011), 'The Anatomy of Bergsonian Temporal Ontology by McTaggart's Problematics', presented at the 62nd West Japan Philosophical Association, Kyushu University, Ohashi Campus, 26 November 2011.
Hirai, Y. (2020), 'Bergson's Panqualityism', *Gendai shiso*, 48 (8): 138–51.
Iimori, M. (2020), 'Harman, Whitehead and Panpsychism', *Gendai shiso*, 48 (8): 152–72.
Inaga, M. (2013), 'A Study on the Contemporary Examination of Bergson's *Matter and Memory*: Its Connection with Chalmers' Naturalistic Dualism', The Japanese Philosophical Society's Hayashi Foundation Research Report Paper. http://philosophy-japan.org/download/857/file.pdf (accessed 27 April 2022).
Kaku, M. (2014), *The Future of the Mind: The Scientific Quest to Understand, Enhance, and Empower the Mind*, New York: Doubleday.
Kelly, M.R., ed. (2010), *Bergson and Phenomenology*, Houndmills, Palgrave Macmillan.
Koch, C. (2011), *Consciousness: Confessions of a Romantic Reductionist*, Cambridge, MA: MIT Press.
Kreps, D. (2015), *Bergson, Complexity and Creative Emergence*, Houndmills, Palgrave Macmillan.
Massimini, M. and Tononi, G. (2013), *Nulla Di Più Grande: Dalla veglia al sonno, dal coma al sogno. Il segreto della coscienza e la sua misura*, Milan: Baldini and Castoldi.
Matsuo, Y. (2016), 'Intelligence for Raising the Probability of Survival' in Y. Matsuo (ed.), *What is Artificial Intelligence?* supervised by the Japanese Society for Artificial Intelligence, Tokyo: Kindai-Kagakusha.
Miyake, Y. and Y. Hirai (2018), 'Incorporating the Bergson model into artificial intelligence', in Y. Hirai, H. Fujita and S. Abiko (eds), *Rebooting Bergson's Matter and Memory*, 120–38, Tokyo: Shoshi-shinsui.
Moravec, M. (2019), 'Perpetual Present: Henri Bergson and Atemporal Duration', *European Journal for Philosophy of Religion*, 11 (3): 197–224.
Moravec, M. (2022), 'A Bergsonian Response to McTaggart's Paradox', in Y. Wolf and M. Sinclair (eds), *The Bergsonian Mind*, 417–31, New York and Abingdon: Routledge.
Mullarkey, J., ed. (1999), *The New Bergson*, Manchester: Manchester University Press.
Mutch, A. (2016), 'The Limits of Process: On (Re)reading Henri Bergson', *Organization*, 23 (6): 825–39.
Nakajima, H. (2016), 'Artificial Intelligence as a Constructive Discipline', in Y. Matsuo (ed.), *What is Artificial Intelligence?* supervised by the Japanese Society for Artificial Intelligence, Tokyo: Kindai-Kagakusha.
Nishimura, S. (2014), 'Is Perceptual Experience Temporally Extended?' *The Hikone Ronso*, 400: 120–35.
Nubiola, J. (2014), 'What a Scientific Metaphysics Really is according to C.S. Peirce', *Cognitio*, 15 (2): 349–58.
Nyíri, J.K. (2014), *Meaning and Motoricity: Essays on Image and Time*, Bern: Peter Lang GmbH.
Ohta, K. and M. Oguchi (2014), 'Cognitive Science of Thought and Rationality', in Y. Nobuhara and H. Ohta (eds), *New Philosophy of Mind I: Cognition*, 111–64.
Olma, S. (2007), 'Physical Bergsonism and the Worldliness of Time', *Theory, Culture and Society*, 24 (6): 123–37.

Papanicolaou, A.C. and P.A.Y. Gunter, eds (1987), *Bergson and Modern Thought*, Reading: Harwood Academic Publishers.

Peirce, C.S. (1931–58), *Collected Papers of Charles Sanders Peirce*. Vols. 1–8, C. Hartshorne, P. Weiss, A.W. Burs (eds), Cambridge, MA: Harvard University Press.

Prigogine, I. (1980), *From Being to Becoming: Time and Complexity in the Physical Sciences*, San Francisco: W.H. Freeman.

Riggio, A. (2016), 'Lessons for the Relationship of Philosophy and Science from the Legacy of Henri Bergson', *Social Epistemology*, 30 (2): 213–26.

Ross, D., J. Ladyman and H. Kincaid, eds (2013), *Scientific Metaphysics*, Oxford: Oxford University Press.

Sato, R. (2014), 'The Neural Basis of Visual Consciousness Controversy: Focusing on the Merits and Demerits of the Dissociative Theory and the Apparent Richness of Perceptual Experience', in Y. Nobuhara and H. Ohta (eds), *New Philosophy of Mind II: Consciousness*, 81–130.

Sinclair, M. (2019), *Bergson*, Abingdon: Routledge.

Smolin, L. (2013), *Time Reborn: From the Crisis in Physics to the Future of the Universe*, New York: Houghton Mifflin Harcourt.

Takamura, N. and T. Suzuki (2020), 'Will panpsychism be rebooted?' *Gendai shiso*, 48 (8): 110–23.

Tani, J., Y. Miyake and Y. Hirai (2018), 'Bergson and the future of artificial intelligence', in Y. Hirai, H. Fujita and S. Abiko (eds), *Rebooting Bergson's Matter and Memory*, Tokyo: Shoshi-shinsui, 139–74.

Taniguchi, T. (2014), *Symbolic Emergence Robotics: An Introduction to Mechanisms of Intelligence*, Tokyo: Kodansha.

Taniguchi, T., Y. Hirai, Y. Tsuda and Y. Miyake (2021a/2021b) 'Exploring the Constructability of New Intelligence and Consciousness inspired by a Bergsonian "Time Scale" model [Part 1/Part 2]', *Artificial Intelligence*, 36 (5): 500–11/627–41.

Tononi, G. (2004), 'An Information Integration Theory of Consciousness', *BMC Neuroscience*, 5 (42).

Watanabe, S. (1951), 'Le concept de temps en physique moderne et la durée pure de Bergson', *Revue de Métaphysique et de Morale*, 56 (2): 128–42.

Williams, C. (1998), 'A Bergsonian Approach to A- and B- Time', *Philosophy*, 73 (285): 379–93. https://www.jstor.org/stable/3751989

Wolf, Y. (2021), '"A Memory within Change Itself." Bergson and the Memory Theory of Temporal Experience', *Bergsoniana*, 1, https://doi.org/10.4000/bergsoniana.286

Zahavi, D. (2003), *Husserl's Phenomenology*, Stanford, CA: Stanford University Press.

Zahavi, D., ed. (2012), *The Oxford Handbook of Contemporary Phenomenology*, Oxford: Oxford University Press.

Part One

Memory and Mind

1

A Secret Connection Between the Dualities Actual/Virtual, and Intelligence/Intuition in Bergson's Metaphysics

Paul-Antoine Miquel
University of Toulouse Jean Jaurès

Introduction

In this chapter, we would first like to insist on a fundamental conceptual device that is at work in Bergson's metaphysics of becoming: the duality 'actual/virtual'. Thinking reality for Bergson is possible only by intuition. In other words, he refuses any speculation concerning reality that is not at the same time an experience. However, the experience of reality is certainly not limited to an experience of duration lived by consciousness. Such a philosophical claim would be a complete misunderstanding of Bergson's philosophy (Deleuze 1968; Worms 2006; Miquel 2007; Ansell-Pearson 2018). The duality 'virtual/actual', as a conceptual device set up in *Matter and Memory* (1896) allows us to explain why a metaphysics of becoming requires an enlargement of experience that cannot be limited to the point of view of consciousness. This philosophical scheme is crucial for understanding how we move from the problem of the duration experienced by consciousness to the problem of life present in the universe.

However, the virtual/actual conceptual scheme can only be rendered operational by a deep reform of the concept of intuition. The French philosopher must move from a vision of intuition that opposes intelligence and science to a very different vision through which intuition 'rides' perception and intelligence (Bergson 1934: 87). The consequence is that perception and science have also an absolute access to reality by experience (Bergson 1907: 199; 1934: 71). It means that the experience of reality proposed by perception and science, is not simply empirical. It is an ontological one. We will explain why in this chapter.

1.1 The problem of memory

The duality actual/virtual first allows Bergson to present an innovative theory of memory that is not limited to the associationist approach. First of all, there is a memory that is usually action-oriented, and memory is then reduced to the frequent trace of a previous event, or of the frequent causal relationship between several events that appear in the same order. The characteristic of this memory is that it will automatically engage the same type of behaviour. One can imagine then how such traces of events or relations can be selected by a trial-and-error mechanism, until an optimal result is obtained from an adaptive and biological point of view.

However, deep memory is something other for Bergson. He gives the example of dreams (Bergson 1896). Deep memory shifts consciousness to originally unconscious areas, where souvenirs have no direct adaptive value. Souvenirs are nothing but underlining the creative dimension of memory. Memory does not simply repeat the same mechanisms, but invents new images, by analogy and metonymy, transporting us to the pure 'inextensive and impotent' (Bergson 1896: 156) souvenir. Souvenir is inextensive, because it is virtual, and it is impotent, because it is not action oriented. Memory has a creative value, because it is the mark of a past that 'extends (*prolonge*) into the present' and thus 'encroaches' (*empiète*) on the future (1934: 28). This formula deserves closer attention.

Indeed, the vision of duration proposed by Bergson in his first work (1889) could be reduced to a series of paradoxes, which are simply contrary to common opinion, but which do not pose any real theoretical problems. It is quite true, for example, that the duration experienced by consciousness is succession, so that when the present arises, the old present has disappeared. They cannot therefore coexist on the same line, and the spatialization of time thus misses the very nature of duration, which is summed up in *the non-co-existence* of past and present. Thus presented, past, present and future, are analogue to McTaggart's A-series (1908).

1.1.1 The crucial distinction between 'coexistence' and 'coextensivity'

Yet if we look more closely at duration conceived in this way, we may well accept that the notions of past, present and future contain something paradoxical, but the paradox seems to be lifted if we focus our attention on the before/after relationship, which corresponds to what McTaggart calls B series (McTaggart 1908). The order of time always seems to be respected in the series of events, in the sense that the after cannot be prior to the before. If we try to compare these events, we end up with paradoxes, but as long as we are content to order them, the before/after relation seems to be preserved. Of course, the idea that there can be a causal chain of events between physical phenomena, in the sense that what happens now depends on events that happened before, has its basis in this conception of time. However, such a relational vision of time, already proposed, not only by McTaggart, but also in France by Berthelot (1911), as a critical objection to the Bergsonian conception of duration, does not solve the problem of creative memory. In pure memory, the past no longer appears as 'coexistent', but as 'coextensive' with consciousness (Bergson 1896: 168).

To be more explicit, the 'coexistence' of two commensurable segments is a property of spatialized time, and the 'coextension' of the past in the present is a property of internal duration. It means of course, that these two words are not synonymous. The past is *in* the present, as a *virtual* dimension of it. The previous sound of a melody is always here in my consciousness. It doesn't disappear in the present, even if it was also in the past. This internalization of the past into the present is nothing but the pure motion by which consciousness recognizes its spiritual power in itself and in its own operations. This virtual dimension is therefore no longer outside the present. There is no need to look for it elsewhere in another supranatural world.

However, by this virtual dimension of duration, the before/after relation of order is challenged, because on the one hand, the past as an ancient event appears before the present, but on the other, the past is *coming back* in the present, as internalized by memory. The physical and material order of succession between events starts to be challenged by a new virtual dimension, which appears in memory, and which is the signature of consciousness as a spiritual entity. As developed by Bergson in *Creative Mind*, spirituality is nothing but some kind of non-trivial recursive process (see Hui 2019) through which consciousness 'draws from itself more than it has' (Bergson 1934: 31).

Characterized as such, consciousness is thrown out of itself because it will continuously internalize what it was, and so that its future depends on this internalization, and not only from the causal relation between past and present events. By making its own what it has been, consciousness is thus projected beyond what it is. As a creative reality, it is related to itself, by some kind of creative circle (Hwang 2019) through which 'it expands and overcome itself' (Bergson 1907: 53) in a future that is also anticipated in the present, as a tendency. The future appears neither simply as a future event, nor as an end or a final cause, but *as its own future*, influenced by *its own history*. As such, cannot be predicted by any classical physical law, or any classical mathematical equation. Looking carefully at the text, this point already appears in *Matter and Memory* (Bergson 1896: 250), even if this book is more focused on the past, than on the future.

1.1.2 *The duality actual/virtual challenges the principle of excluded-middle*

Consciousness, as a virtual and spiritual entity, draws from itself more than it has (Bergson 1934: 31) for a second crucial reason already developed in *Matter and Memory*: essentially 'virtual', memory is then that movement by which the past 'expands into the present image thus emerging from obscurity onto the light of the day' (Bergson 1896: 150).

As a spiritual entity, consciousness will never be in perfect adequation with itself because consciousness is not a substance, it is a movement. It moves in a kind of immobile way because it is always both behind and ahead of itself. It is not simply because of its memory and its power to anticipate the future, but also and because it cannot be virtual without having the power to emerge in the present. In the same way that there is no shadow without light, there can be no virtual without actuality. *Thus, virtuality is otherness.*

As a past, the virtual is a difference, it is not a substance. It is the internal and qualitative difference between my present and my past. The past as such is constantly other, and it is through this otherness that duration as consciousness is made. It is nothing ready-made, because it is *the relation* between *its being* and *its acting* that *defines its being*, which is a fundamental characteristic of all philosophy of becoming (see also James 1907; Whitehead 1929 or Simondon 1995). Consciousness, then is not a substance, it is a process.

How should we think about the present then? The present is actual. It is another dimension of time. However, just as it is not possible to understand the virtual without the actual, it is not possible to understand the actual as a pure nothing. *Bergson's logic does not respect the principle of the excluded middle, because what is not virtual is not nothing.*[1] *Rephrased in a more symbolic language:*

(1) If A, not A is not nothing.

The actual *is not a contrary*, since it is not independent of the virtual. But it is also *not a contradictory*, because it is a negation which cannot be understood as a nothing. Indeed, negation must be conceived as no longer being radically separated from affirmation, since the present, as *actual is other the otherness*. Let us emphasize this point: my present is not a dimension of time, as width would be a dimension of space. Length, width and depth can be considered as independent variables in a given space. On the contrary, the present and the past are ecstatic dimensions of time. Ecstatic dimensions, unlike variables, are penetrable entities, that communicate with each other. They are intertwined. They cannot be separated.

The last point remains to be elucidated: why consciousness, as a spiritual entity, only expands and overcomes itself while at the same time lagging behind itself, precisely because my past, as a penetrable entity, depends on my present from which it cannot be dissociated, but my present ties me precisely to space and matter. The cross-section of the present is only a plane of materiality, the plane of coexistence, which is the plane P that the cone of memory touches with its tip S in Bergson's diagram (Bergson 1896: 169). Consciousness can therefore only project itself beyond itself, as a radically non-spatial entity, in a plane of spatiality. What is not itself is thus part of it, so that it can only be related to itself by going beyond itself. This is why it always draws from itself more than it has. This is the most profound paradox of a true philosophy of spirituality, and there is no longer any contradiction behind this paradox, which offends common opinion.

The tip S symbolizes my present, in a plane of spatiality P in which an infinity of other presents coexist: this relation of coexistence can be grasped by a spatial scheme, understood by Bergson as a Euclidian one. Since space is defined precisely as what succession is not, coexistence means simultaneity. However, nothing in this way of thinking would be opposed to the idea that coexistence and simultaneity *are not necessarily synonymous*. The crucial point is to characterize matter as coexistence and actuality, without denying, that actuality is also *a penetrable entity*.

1.1.3 Pure perception, as an ontological experience

The duality virtual/actual is in connection with a dual experience of reality. Following the first one, the past of consciousness is also my past, experienced from *the point of view of consciousness*, but at the same time, the present is also lived as my present, understood *from the point of view of the material universe in which my present is included.* Thus, the *experience* of the presence of my mind, and of my brain into the material universe, is nothing but 'pure perception' (Bergson 1896: 31). By pure perception, consciousness is open to the coexistence of images, which cannot fit with the coextensivity of memory. Objects perceived are not only images. For me, they are not only representations because I believe immediately that they exist independently of the way I am thinking of them. I see the table here, and I know by some kind of 'common sense' intuition, that this table exists. Why is it so? Because *in my perception,* the table, the tree, and finally the garden, *are images related to themselves.*

As far as the ego is concerned, it does not exist, as such, in my perception. When I perceive, I also perceive my body. Thus, my body is an image in this world of images, but to mean that, perception comes from my body and my sense awareness is nothing other than accepting that I perceive material images in the world of images. Through this experience, I do not see the images as correlated with my subjective consciousness. *I see, on the contrary, my perception as an image included in another world, to which my consciousness is also open:*

> You may say my body is matter, or that it is an image: the word is of no importance. If it is matter, it is a part of the material world; and the material world, consequently, exists around it and without it.
>
> Bergson 1896: 14

This is an experience of contemporaneity, through which *I am going out* the immanent world of the duration lived by my consciousness. More precisely, this new experience means that I see the table, the tree and the garden as being linked to the same objective world of images, and that I see these images as also linked to my body, as a subjective centre of action and decision. The ideal perception is nothing more than the pure limit between these two perspectives: the world of images in relation to themselves and the world of images in biological relation to my body and my consciousness.

Finally, to understand the contemporaneity of the material world, I must let both perspectives emerge in my mind. I cannot remain a prisoner of a subjective point of view, since by my perception I see things 'in themselves' (Bergson 1896: 2). Such an assertion comes from the fact that contemporaneity, as the coexistence of material images, is another dimension of duration, *already present in my mind*: the dimension of the present. There is no duration without actuality, and therefore I have no experience of duration lived by my consciousness without at the same time experiencing what is outside my consciousness and uncorrelated with it, namely the experience of my body, and then that of the relationship between my body and the other bodies of the universe. This experience shows the locality of our temporal consciousness in the whole universe.

1.2 What is at stake in the material world?

When we listen to a melody, the memory of the sounds and their harmony is experienced in an immanent way in the world of our thoughts, which is the deep meaning of the term 'coextensive'. However, the virtual memory of these sounds could not be actualized, if it did not return in a present, which is also that of my body. Through this experience, I no longer see the things in me, but rather I am in the things I see. In *Creative Evolution*, Bergson will give a name to this enlarged experience. He will call it the experience 'of the existence in general' (Bergson 1911: 7). The experience of the existence in general is not a simple psychological experience. It has nothing to do with introspection, but it is indeed an experience. Perception places us outside ourselves, but 'it makes us touch the reality of the object by an immediate intuition' (Bergson 1896: 79).

By this experience, we participate in the material dimension of the universe, through which my present is contemporary with an infinity of other presents connected with it. It is the present of 'this set of images' of which Bergson speaks. Two questions then arise, and we will address them to finish this chapter.

The first is to know if there can exist, outside my present, other forms of spiritual consciousness similar to mine? We will answer this question by a simple speculative argument. The couple actual/virtual by definition supposes that it is not possible to disconnect the actual from the virtual. Let's remember that the actual and the virtual, in other words, the present on one side, the past and the future on the other, are not dimensions that we could disentangle from each other, like the axes of a Euclidean frame of reference. They are temporal ecstasies. There is no virtual without actual, by hypothesis, but it is thus reciprocally that there cannot be actual without virtual. Each material point of the universe, in the simple fact that it is linked at the same time with other points, by a relation of continuity which is proper to what Bergson names 'the concrete extension' (*étendue concrète*, Bergson 1896: 208, 247) is for this reason more than a simple point. If it is a particle, it is at the same time a wave. If it is discontinuous, it cannot however, by certain aspects of its being, not be at the same time continuous. It participates as such of a consciousness (Bergson 1896: 264; Dolbeault 2018), although Bergson names it a sleeping memory. There is thus no reason to refuse that in the present of another material body than mine, such a memory can also take a less repetitive form, and more creative than that of a simple material point. However, speculative arguments are never enough for Bergson. The only real question that is raised is whether we can experience spiritual forms other than our own in the material world. If so, what form of experience is it? To answer, we cannot proceed by simple introspection, since it is the material world that is at stake, and not what is subjectively experienced in our consciousness.

The second is would perception be the unique mental activity by which the material world can be experienced? The answer is no. Perception shares with science the ability to put us in contact with the material world:

> Now no philosophical doctrine denies that the same images can enter at the same time into two distinct systems, one belonging to *science*, wherein each image, related only to itself, possesses an absolute value; and the other, the world of cons*ciousness*, wherein all the images depend on a central image, our body, the variations of which they follow.
>
> Bergson 1896: 21

More completely than perception, but in a gesture which prolongs it, science touches the reality of the matter with the spatial scheme by introducing a point of view of the system distinct from the point of view of the observer. This new point of view makes it possible to objectify and quantify physical qualities, to identify physical laws, and to make predictions. In doing so, science involves an ontology of the actual, which is the very one through which our thought becomes bound up with the material world.

Then, the second characteristic of science is to put the accent on an ontology of the actual, which will thus privilege invariance. This is why duration has no influence on the material systems analysed by science. It is not because its action is artificial. It is rather because it emphasizes in reality only its actual dimension. In other words, science is not made to understand the spiritual dimension of reality.

However, insofar as it is not possible to understand the actual without the virtual, the new forms that the interrogation of reality will take will depend on the point of view of science. This is, in our opinion, the most innovative legacy of Bergson's metaphysics. Even if it is the task of philosophy to question the spiritual, it cannot do it alone, *a fortiori*, when it is a matter of tracking down the presence of this spiritual dimension, not simply in our consciousness, but also in the material world. *The ontological experience only really takes shape when it becomes dual.* It is no longer possible for the philosopher to question reality alone. They must do it in dialogue with the scientist. Metaphysical thought can only emerge in and through this demanding dialogue. This theoretical position clearly distinguishes Bergson's philosophy from the phenomenological ontology developed by Heidegger, since the question of the real is not reserved to the philosopher, and since the question of the real is also the question of the becoming, and not the question of the being.

Our first question remains: can we detect the presence in the material world of spiritual entities able to 'draw from themselves more than they have'? Above all, what kind of experience do we need to do this? The answer to the question we are asking is yes: there is indeed a force in the material world that is capable of drawing more from itself than it has. This is life.

Like human consciousness, life is always both ahead and behind itself, precisely because it is also material. Life is tendency. It is finite, Bergson tells us again (Bergson 1907: 128, 254). It is not Final Cause, or Substance. Its limit is none other than its actual dimension. Life therefore materializes not only in bodies, but also in physical principles. Life does not depend simply on matter, but also on physical laws. There is no life without waste, without entropy (see Miquel 2007; Di Frisco, 2015). To detect light, the plant already needs to expend energy, but it is this very energy that enables it to set up detection devices, thanks to which it can grow. To detect food, the animal needs to move, but it is this new ability that will give it the instinct that the plant does not have. Instead of simply applying organization, like the plant, the animal expends action, but a human being is spending even more, because they can waste their thoughts. Following Bergson, it is only in this way that the human intellectual function is created. It is therefore not enough to say that life makes itself, in the sense that, like human consciousness, it exceeds itself. Life 'makes itself' through an entropic flow of 'unmaking itself' (Bergson 1907: 248). The key to biological organization is therefore not the concept of maintenance, homeostasis or regulation, as was the case with Claude

Bernard, and as will again be the case with Varela and Maturana (1974). For Bergson, every biological system tends to recompose itself, precisely because it can only build by spending. It can only create by destroying. Wherever life is registered, there is a dynamic polarization of the living. The living being thus remains in constant inadequacy with itself. This is the secret of its organization, and it is what makes it capable of being pushed by itself beyond the present, towards a future that is also its own, and no longer a fatal destiny.

A final question then arises. Could such a concept ever be used to construct a scientific theory of life? This is not Bergson's opinion. According to him, intelligence cannot understand this spiritual movement which is present on a large scale in life, and on a small scale in human consciousness. Life only closes in on itself by remaining permanently inadequate to itself. This is the key to the concept of creative evolution. Only in this way can it be described as something that draws more from itself than it has. He does not have the idea that this conceptual scheme might express only a particular form of closure and that formal language might have some resources to express it (Miquel and Hwang 2016).

He remains stuck on an argument that seems central to us: matter is not becoming. Matter is only the actual dimension of becoming. Any attempt to reduce becoming to matter will also tend to reduce becoming to being. Being for Bergson, is only a property of becoming, and concerning Being there is no need to look for more. If we want to focus on Becoming itself, we immediately understand that Becoming is what it does. It is nothing but the relation between what it is and what it does that characterizes what it is. The concept of Process is therefore the true fundamental concept in Bergson's metaphysics.

When science analyses matter, it never looks at the purely actual dimension of becoming. When science analyses the movement of the planet, it always finds the infinitesimal and the mathematical continuum under the denumerable, when it tracks down the electron, it always finds the quantum under the particle. Entropy itself, in statistical mechanics, is nothing more than a tendency: the same probability distribution that returns indefinitely in an ergodic system. However, the fact that matter is always partially continuous and virtual does not change the fact that the scientist can only analyse it by reducing, as far as possible, the virtual to the actual, the time to the space, the variation to the invariant.

There are therefore two opposing movements in human thought, and a true philosopher of becoming must accept this. On the one hand, intelligence seeks structures, on the other, intuition seeks what in reality goes beyond the structures that intelligence has found. Intuition therefore no longer turns away from intelligence, it overlaps it: it is only in this way that it can sometimes overcome it.

Note

1 The principle of excluded middle states at the epistemic level, that for any proposition, either that proposition is true or its negation is true, in other words: $\neg\neg p \Rightarrow p$, and at the ontological level that, if there is A, then not A cannot be.

References

Ansell-Pearson, K. (2018), *Bergson: Thinking Beyond the Human Condition*, London: Bloomsbury Academic.

Bergson, H. (1889), *Essai sur les données immédiates de la conscience*, Paris: Alcan. *Time and Free Will* (1910), trans. F.L. Pogson, London: George Allen and Unwin.

Bergson, H. (1896), *Matière et mémoire,* Paris, Alcan. *Matter and Memory* (1911), trans. N.M. Paul and W.S. Palmer, London: George Allen and Unwin.

Bergson, H. (1907), *L'évolution créatrice,* Paris, Alcan. *Creative Evolution* (1911), trans. A. Mitchell, London.

Bergson, H. (1934), *La pensée et le mouvant,* Paris, Alcan. *Mind-Energy* (1920), trans. H. Wildon Carr, New York: Henry Holt and Co.

Berthelot, R. (1911), *Un Romantisme utilitaire, étude sur le mouvement pragmatiste. 2. Le pragmatisme Bergson*, Paris: Alcan.

Deleuze, G. (1966), *Le Bergsonisme*, Paris, Presses universitaires de France.

Di Frisco J. (2015), 'Élan Vital Revisited: Bergson and the Thermodynamic Paradigm', *The Southern Journal of Philosophy*, no. 53 (1).

Hui, Yuk (2019), *Recursivity and Contingency*, London: Media Philosophy.

Hwang, Su-Young (2017), 'La figure du cercle vertueux comme aspect essentiel de la causalité créatrice. Une lecture critique de Bergson', *Dialogue*: 56.

James, W. (1907), *Pragmatism: A New Name for Some Old Ways of Thinking*, New York: Longmans, Green.

McTaggart, J.E. (1908), 'The Unreality of Time', *Mind* XVII. 4: 457–74.

Miquel, P.A. (2007), *Bergson ou l'imagination métaphysique*, Paris: Kimé.

Miquel P.A. and Hwang Su-Young (2016), 'From Physical to Biological Individuation', *Progress in Biophysics and Molecular Biology*, 122 (1): 51–7.

Simondon, G. (1995), *L'individu et sa genèse physico-biologique*, Paris: Million.

Varela, F. H. Maturana, H.R. and R. Uribe (1974), 'Autopoiesis: The Organization of Living Systems, Its Characterization and a Model', *Biosystems* 5: 187–96.

Whitehead, A.N. (1929), *Process and Reality*, New York: The Free Press.

Worms, F. (2006), *Bergson ou les deux sens de la vie*, Paris: Presses universitaires de France.

2

Bergson and the Rise of the 'Sciences of Memory'

Takeshi Miyake
University of Kagawa

1 Introduction

In *Rewriting the Soul: Multiple Personality and the Sciences of Memory* (1995), Ian Hacking argues that the 'sciences of memory' appeared in the late nineteenth century and that these sciences replaced the soul with knowledge about memory (Hacking 1995: 198). In other words, the soul was simultaneously 'scientized' and secularized (Hacking 1995: 251) by these sciences. Hacking considers that the structure of the sciences of memory came into being during 1874–86. This was ten years before Henri Bergson published *Matter and Memory,* and Hacking mentions Bergson only once in *Rewriting the Soul*. So, how do these sciences of memory emerge in *Matter and Memory*? What influence did they have on Bergson's argument in *Matter and Memory*? These are the problems we want to discuss. In this chapter we focus on the second chapter of *Matter and Memory,* in which Bergson refers to many scientific works.

Hacking applies Michel Foucault's method of analysis in his argument and lists three sciences of memory.[1] These are (a) the neurological studies of the location of different types of memory; (b) experimental studies of recall; and (c) what might be called the psychodynamics of memory (including the positive psychology of Ribot) (Hacking 1995: 199).[2] We will examine how the rise of these sciences of memory is related to the construction of the theories in *Matter and Memory*. First, we deal with the experimental studies of recall, and then the other two sciences in Hacking's list, because it is convenient for us to develop our study in this way.

2 Experimental psychology and *Matter and Memory*

It is well known that Bergson suffered from insomnia when he wrote *Matter and Memory*, because he had to spend five years reading an enormous amount of research on aphasia. Indeed, many reference sources are cited in the second chapter of *Matter and Memory*: around seventy researchers are mentioned. Bergson takes into consideration various kinds of research: philosophy, psychology, psychiatry, neurology, pathology, etc., and consequently it is difficult to understand their relationship. In this section, we will identify and describe the works of experimental psychology that are mentioned in *Matter and Memory*.

Hacking claims that Hermann Ebbinghaus (1850–1909) established '(b) the experimental studies of recall', explaining that Ebbinghaus differed from Fechner, the founder of psychophysics, in that he introduced the statistical method into experimental psychology: 'The critical feature of Ebbinghaus's work was that he instituted statistical treatment of data' (Hacking 1995: 204). This claim is justified because the method of statistics had a revolutionary influence on the sciences at that time. That is demonstrated by Canguilhem (*The Normal and the Pathological*), Foucault and Hacking himself (*The Taming of Chance*).

Although the works of Ebbinghaus are not referred to in *Matter and Memory*, we can see his influence on other researchers quoted by Bergson in the second chapter of the book. For example, Bergson refers to the experiments conducted by Georg Elias Müller (1850–1934) and William George Smith (1866–1918) (MM: 99–100). At the beginning of his paper, G.E. Müller discusses Ebbinghaus's book with F. Schumann (Müller and Schumann 1894: 6), while Smith makes reference to the papers of Ebbinghaus and Müller as preceding studies (Smith 1895: 47, 58, 64, 72).

When Bergson invokes the results of their experiments, he examines the connection between movements (habit) and images (voluntary recollection).[3] He concentrates on a case where the memory-image intervenes latently in the formation of a habit. For instance, in the experiment of Smith, the subjects are given ten seconds to memorize twelve letters arranged in three lines (Figure 2.1), and then, after a pause of two seconds, they are required to recall the letters.

During these operations, the subjects are required perform simultaneously one of the following actions: 1. a simple addition sum, viz., that involved in repeating the series 2, 4, 6, 8, . . .; 2. repeat a simple syllable, such as *la*; 3. tap the table with their forefinger; 4. do nothing. The purpose of this experiment is to extract data showing how these operations prevent the subjects from remembering. In some cases, when the subject endeavors to repeat the first letters, they have simply disappeared.[4] According

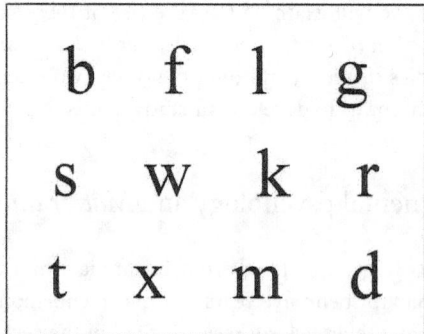

Figure 2.1 An example of the card used in Smith's experiment. (Smith 1895).

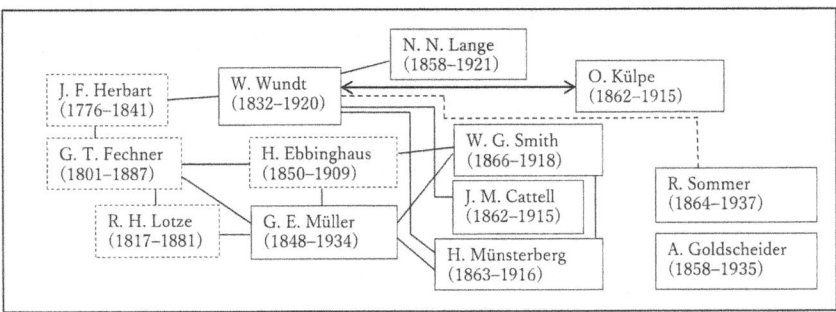

Figure 2.2 Experimental psychologists in *Matter and Memory*. © Takeshi Miyake.

to Bergson, we have to make use of a fugitive image in order to form a habit and this fugitive image (a representation of an ensemble) suggests the action of pure memory.[5] In this way, Bergson invokes 'the experimental studies of recall' to reveal the subtle relationship between two types of memory. Images work virtually on the formation of a habit, but this image disappears immediately when we make an effort to remember.

In the second chapter of *Matter and Memory*, Bergson refers to other experimental psychologists. Here is a schematic diagram (Figure 2.2) that represents the relationships between them (those bounded by broken lines are not mentioned explicitly in *Matter and Memory*). From this diagram, we can identify many researchers who studied under Wilhelm Wundt: Hugo Münsterberg, who went to the United States on the invitation of William James and who later studied industrial psychology; Nikolai Lange, a Russian experimental psychologist; Oswald Külpe, the founder of the Würzburg School; and James McKeen Cattell, an American psychologist influenced by the statistical method of Francis Galton.

In the second chapter of *Matter and Memory*, there is an important passage where Bergson refers to some of these experimental psychologists together. This is the passage where he examines the subject of *attentive* perception in the first half of Section III, 'Recognition and Attention' (MM: 125). Here, Bergson invokes the results of the experiments of Münsterberg and Külpe. He says that 'any memory-image that is capable of interpreting our actual perception inserts itself so thoroughly into it that we are no longer able to discern what is perception and what is memory' (MM: 125). He then quotes the experiments of Goldscheider and Cattell and argues that rapid reading is a work of divination and that our memory-images are projected onto the perceptions of the letters.[6]

In other words, Bergson is concerned with these experiments when he deliberates on attentive perception, which is a mixture of memory-images and perceptions. In fact, he is also interested in the experiments of others – Lange, Wundt, etc. – in the context of answering the question 'What is an attention?' The above-mentioned experiments of Müller and Smith deal with the relationship between recollection and disturbed attention. When we analyse the context in which Bergson considers the outcome of experimental psychology carefully, he discusses almost always the theme of attention (in relation to memory-image and movement). We will think about the meaning of this when we treat the other two sciences of memory.

3 The localization of memories and *Matter and Memory*

According to Kurt Goldstein, about 2,300 research papers on aphasia were published between 1874 and 1907 (Goldstein 1910: 4). There is no doubt that of the three sciences of memory, Bergson makes the most references to 'the neurological studies of the location of different types of memory'.

Hacking regards the founder of this field of study to be Paul Broca (1824–80), but we should examine this point in detail. Israel Rosenfield, the author of *The Invention of Memory* (1988), says that 'one scientist who suggested that not all memories are necessarily fixed, Paul Broca, was the first one to convince the medical world that brain function is localized' (Rosenfield 1988: 16). It is true that Broca was the first to establish the localization theory of brain function, but it is not clear that he entirely supported the localization of memories too.[7]

If not Broca, who is the founder of the localization theory of the memories? In our opinion, it is Carl Wernicke (1848–1905), who proposed the localization of the memories while correlating it with research into aphasia. Moreover, it is Theodor Meynert (1833–92) to whom Wernicke owes his idea that memory-images are accumulated in the cerebral cortex. 'According to Meynert's calculation, there are 600 million cells in the cerebral cortex, which offer a sufficient number of storerooms where the countless impressions that come from the outside world can be stored up. [...] The cerebral cortex is full of the residues of excitations transmitted. We name this residue memory-image and distinguish it from the impression of sensation itself' (Wernicke 1874: 5, translated by the author). In this way, Wernicke supposes that the memory-images of an articulation are stored in the motor speech centre (Broca's area) and that those of hearing are stored in the sensory speech centre (Wernicke's area).

Ludwig Lichtheim (1845–1928) developed his schema of aphasia from the ideas of Wernicke. To the centre of motor images (M) and the centre of auditory images (A), he adds the centre of visual representation (O), the centre for writing (E), and the centre of concepts (B) (Figure 2.3), although he thinks that B is not localized, replacing B with $B^1, B^2, B^3, B^4 \ldots$ (Figure 2.4).

As pointed out in the second chapter of *Matter and Memory* (MM: 156–7), many scientists (Charcot (Figure 2.5), Broadbent (Figure 2.6), Kussmaul (Figure 2.7), Magnan (Figure 2.8), Moeli (Figure 2.9), Wysman (Figure 2.10), Freud, etc.) have produced various schemata of aphasia. We do not have the space to describe them all, but we cannot understand how Bergson creates the concept of the motor diagram (*le schème moteur*) without taking account of these schemata.

It is significant that Bergson elaborates his concepts of the motor diagram by working on the problem of word deafness and *echolalia* (this is pointed out by the

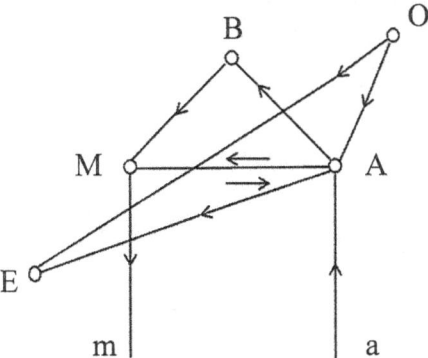

Figure 2.3 Schema of Lichitheim (Lichtheim 1885: 443).

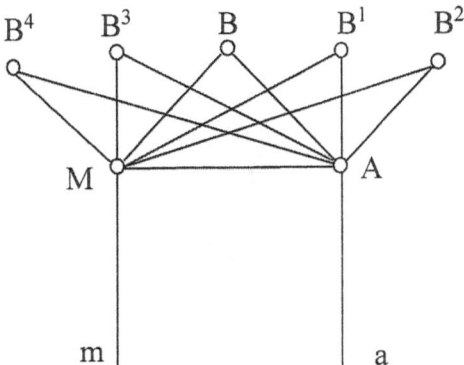

Figure 2.4 Schema of Lichitheim (Lichtheim 1885: 478).

editors in a note in the critical edition of *Matière et memoire*).[8] In this argument (MM: 140–1), he refers to Lichtheim, Charcot's case report, Henry Charton Bastian (1837–1915), Adolf Kussmaul (1822–1902) and so on.[9] At the beginning of *On Aphasia*, Freud says that they are 'the best brains of German and foreign neurology' (Freud 1953: 1).

Bergson compares Bastian with Arnaud and Spamer. Bastian regards the motor accompaniment of speech as a voluntary action, whereas Arnaud and Spamer regard it as automatic and involuntary (Bergson 2012: 365); Bergson estimates that 'the truth appears to lie between these two hypotheses' (MM: 141). There is, in the motor diagram, 'more … than absolutely mechanical actions but less than an appeal to voluntary memory' (MM: 141).

The fact that Bergson does not approach the problem from an anatomical perspective but a functional one is crucial to him compromising between the two

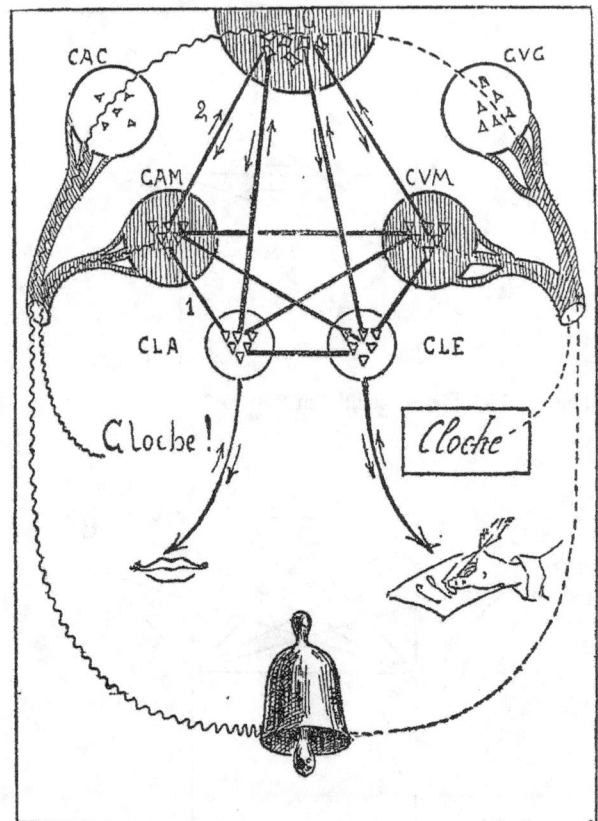

Figure 2.5 Schema of Charcot (Bernard 1885: 45).

hypotheses (this is why he did not visualize his *schème moteur*). While Bastian, Arnaud and Spamer devote themselves to the anatomical aspects (the structure of the mechanism), that is to say, the relationship between impressions and their residues (memories), as well as the relationship between the auditory centre and the motor centre, Bergson focuses on the function of the mechanism: the movement that accompanies a speech we hear articulates that speech. He says that 'the diagram, by means of which we divide up the speech we hear, indicates only its salient outlines. It is to speech itself what the rough sketch is to the finished picture' (MM: 138–9).

In this way, the position of Bergson is closer to holism than to localization theory. Holism explains aphasia by a gradual regression of function, whereas localization theory is based on an anatomy that fixes the seat of aphasia. The former is supported by Broca, Meynert, Wernicke, Lichtheim, Munk, etc., while the latter is held by Jackson, Bastian, Kussmaul, Freud and Sommer, among others. However, the point of this argument does not perfectly coincide with that about the localization of memory-images. For example, Bastian localizes memory in the brain, but he is still regarded as

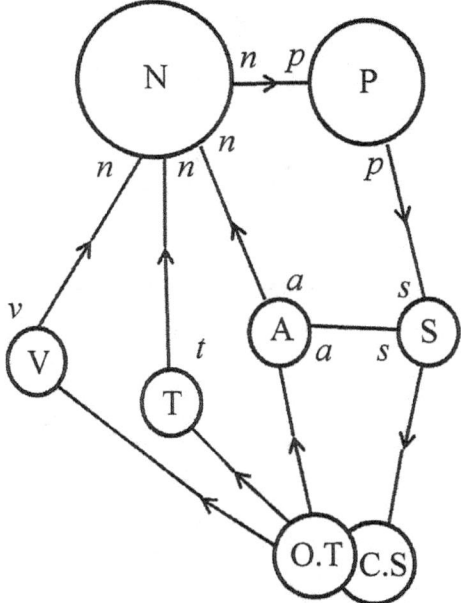

Figure 2.6 Schema of Broadbent (Broadbent 1879: 495).

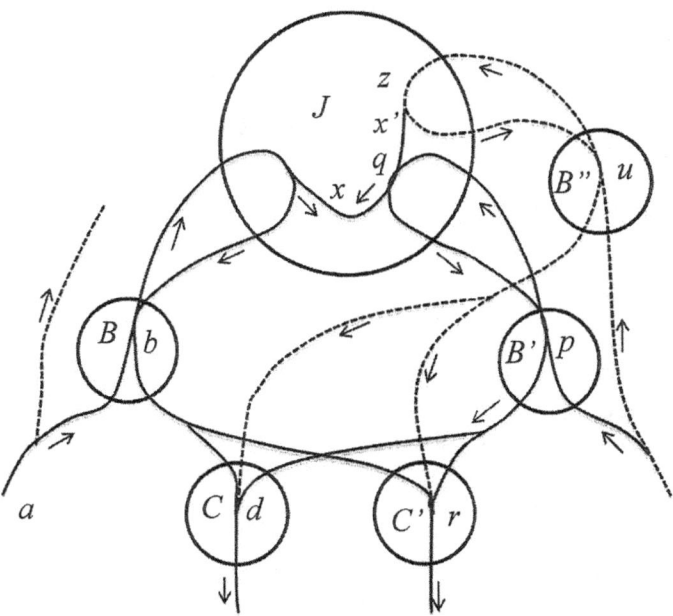

Figure 2.7 Schema of Kussmaul (Kussmaul, translated 1884: 234).

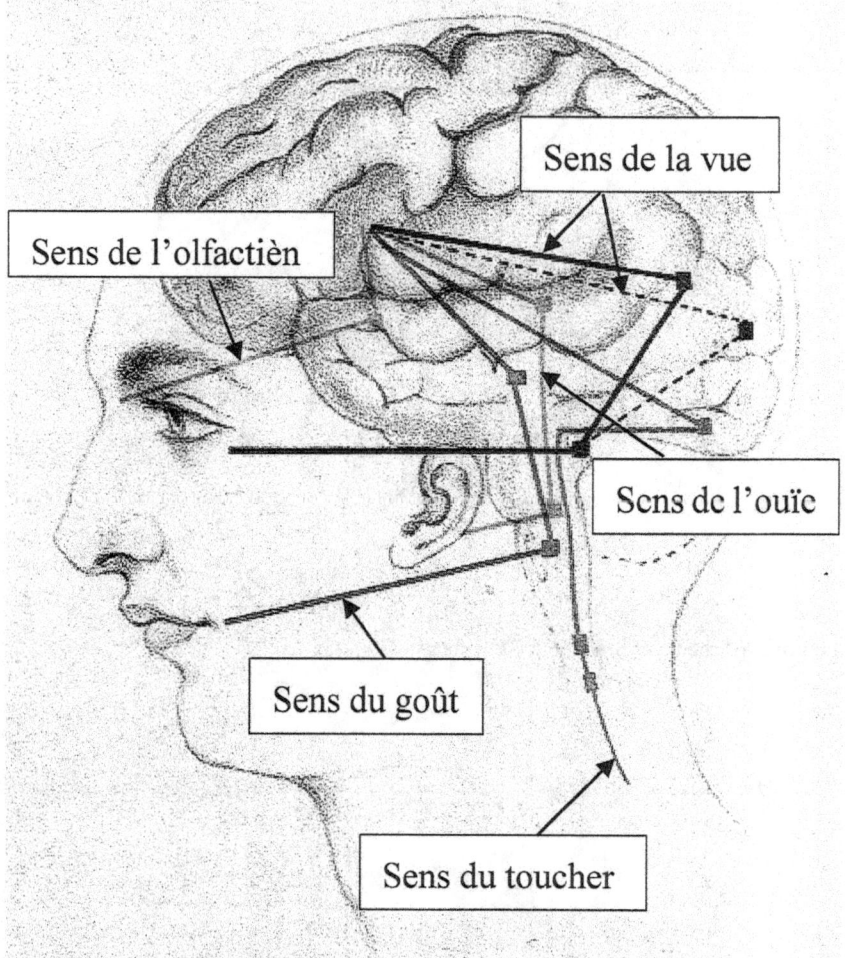

Figure 2.8 Schema of Magnan (Skwortzoff 1881, Planche I) captions added by the quoter.

a holist. We should note that Bergson dedicates himself to discussing the localization of memory-images.

From the members of second group (holists), Bergson's view is most similar to those of Sigmund Freud (1856–1939) and Robert Sommer (1864–1937). We have several reasons to take note of Sommer,[10] but his thought is explained to some extent in the notes of the critical edition of *Matter and Memory* (Bergson 2012: 333–4, 374, 378–9). Therefore, we will examine Freud's *On Aphasia* (1891) because it is difficult to understand at first glance the influence that this book has on the arguments in *Matter and Memory*. However, we can see something in common between Bergson and Freud. The following are some of the similarities: (1) a rejection of the localization of

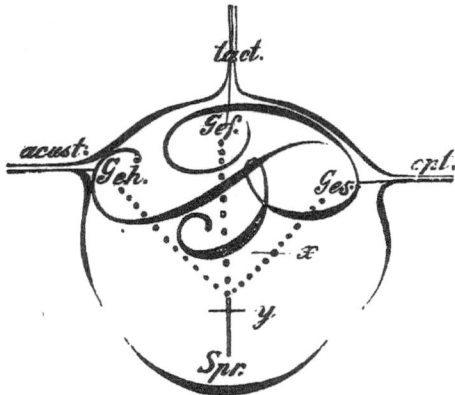

Figure 2.9 Schema of Moeli (Moeli 1890: 380).

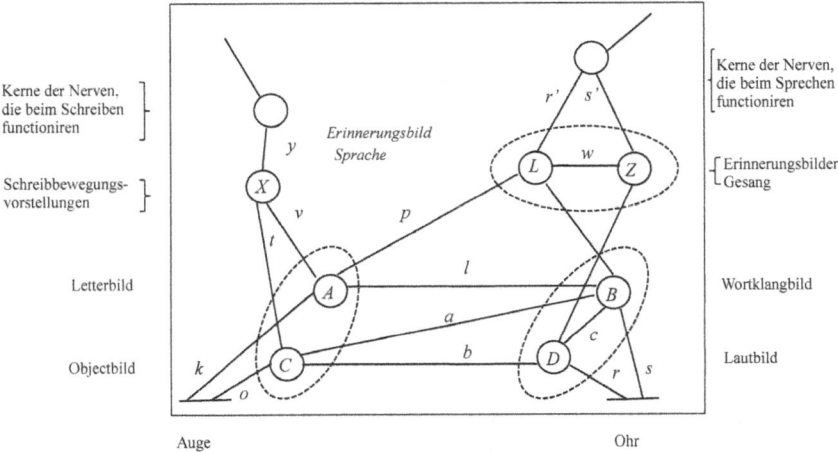

Figure 2.10 Schema of Wysman (Wysman, 1891: 37).

memories; (2) their explanations of aphasia; (3) their ideas about centres; and (4) their criticisms of the localization of memories.

1. Freud definitively refuses the localization of memory-images. He says that '[Meynert's] theory has been evolved that the speech apparatus consists of distinct cortical center; their cells are supposed to contain the word images ... One may first of all raise the question as to whether such an assumption is at all correct, and even permissible. I do not believe it to be' (Freud 1953: 54).[11]
2. Invoking the ideas of John Hughlings Jackson (1835–1911) and those of Bastian, Freud explains aphasia not by localization, but by a functional disorder. Meynert

supposes that an impression is directly brought to the cerebral cortex to be stored up as a memory-image, but Freud thinks that before an individual element (an impression or stimulation) comes to the cortex, it is combined with other elements many times, according to its function.[12] Therefore, he says, 'topographic relations are maintained only as long as they fit in with the certain claims of function' (Freud 1953: 53). This view of Freud is close to Bergson's that the brain is not a storehouse of memory-images but of motor mechanisms (habits).

3. Wernicke distinguishes the centres where images are accumulated from the pathways where images are associated, but Freud does not admit this function of the centre, so he does not need to make a distinction between them. He declares that '[W]e have disposed of an important reason for differentiating between [the] centers and pathways of speech' (Freud 1953: 57). As a result, he identifies the centre with the pathway and then changes the meaning of the concept of the centre. He says that 'the destruction of a so-called center comes about only through simultaneous interruption of several fiber tracts' (Freud 1953: 17). Bergson conceives of the centre in this way, from his knowledge of neurohistology (Charles Pupin), mentioned in the third chapter of *Matter and Memory*.[13]

4. The resemblance between Bergson and Freud is also to be found in the way they criticize the localization of memory-images. Bergson points out that such a view falls into a contradiction between psychology and pathology (or physiology).[14] 'For notice the strange contradiction to which this theory is led by psychological analysis on the one hand, by pathological facts, on the other hand'.

MM: 160–1

Freud criticizes localization theory similarly. It regards the centre as a storehouse of images and the pathway as the place where the images are associated. He argues, however, that '"Perception" and "association" are terms by which we describe different aspects of the same process.... [so] we cannot possibly have a separate cortical localization for each' (Freud 1953: 57).

Both object that memory-images should not be treated as things, but as a process.[15] It is interesting that Bergson criticizes localization theory several pages after he refers to Freud. We can therefore surmise that Freud's argument influenced Bergson. Needless to say, there are differences between Freud and Bergson. It is essential for Bergson to discriminate between two forms of memory, but Freud has no notion of any such discrimination. Freud is attached to the anatomical schema (Figure 2.11), whereas Bergson clarifies the dynamic function of schema. In addition, Freud supports parallelism, which Bergson does not accept.

In summary, it is certain that Broca was the founder of the localization theory of brain function, but it was Wernicke who established the theory of the localization of memories. Subsequently, various schemata of aphasia were invented and are inseparable from the creation of the 'motor diagram' (*le schème moteur*), the most important concept in the second chapter of *Matter and Memory*. When we trace the genealogy of criticism of localization, we can go back to Jackson from Freud, and we can see some

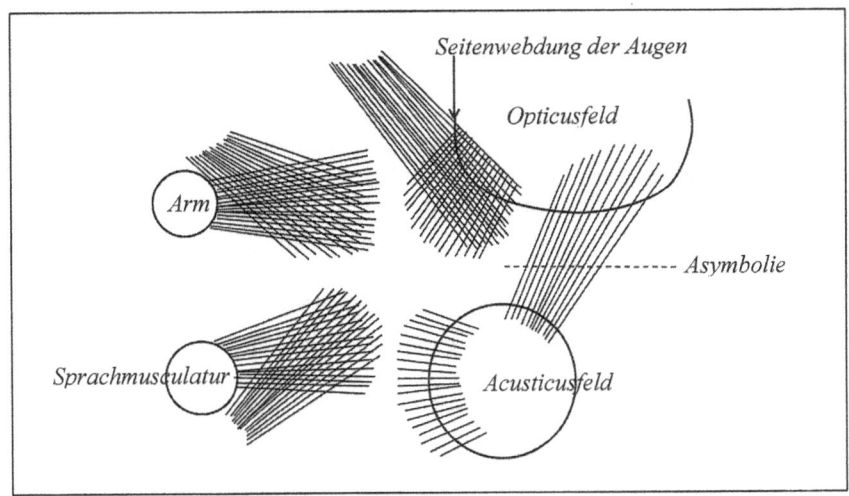

Figure 2.11 Schema of Freud (Freud 1891: 83).

similarities between Bergson and Freud in terms of their objections to the localization of memory-images.

4 The psychodynamics of Ribot and *Matter and Memory*

Of the three sciences of memory, Hacking pays most attention to psychodynamics because it is deeply connected to the themes of scientizing and secularizing the soul. In the second chapter of *Matter and Memory*, we can enumerate the following researchers related to psychodynamics: Spencer, Jackson,[16] Robertson, Kussmaul, Charcot, Féré, Ballet, Bernheim, Ribot, Pierre Janet and Freud[17] among others.[18]

Hacking considers Théodule Ribot (1839–1916) to be the originator of psychodynamics,[19] but he argues that the psychodynamics of memory can never be entirely separated from the work of Freud and that it is 'the study of memory in terms of observed or conjectured psychological processes and forces' (Hacking 1995: 199). Why does not he choose Freud but Ribot as the founder?

The reason is that Ribot had a strategy for scientizing and secularizing the soul. 'The strategy of Ribot and his positivist colleagues was not to attack religious and philosophical ideas of the soul, but to provide a surrogate for the one aspect of a human being that seemed resistant to science. Instead of studying a unitary *moi*, we should study memory' (Hacking 1995: 207–8).

In the second chapter of *Matter and Memory*, Bergson refers to Ribot with reference to the following:

1. there is no perception that is not prolonged into movement (MM: 111);
2. attention and movement (MM: 121);

3. Ribot's law (the law of regression) (MM: 150); and
4. the contradiction between psychology and localization theory (MM: 160–1).

(1) When discussing recognition, Bergson argues that underlying it is a consciousness of a nascent movement that accompanies any perception. This movement is the essence of the motor diagram, as mentioned above. He invokes here Ribot's idea that there is no perception that is not prolonged into movement, in order to support the conception that all images are movements.[20] Bergson agrees with Ribot's view that the nature of perception is movement.

(2) Next, Bergson mentions Ribot when he deals with the subject of attention. Ribot counts attention as a movement of arrest, but Bergson thinks that Ribot's view is inadequate. In addition to Ribot's account, Bergson argues for the positive movement of attention.[21] It is interesting that he quotes the research of experimental psychology when he examines the positive aspect of attention. In this way, Bergson identifies two aspects of attention, arguing that Ribot has grasped only its negative aspect.

(3) In the most important of these references to Ribot, Bergson interprets the meaning of Ribot's law (the law of regression): with some types of amnesia, memory of words disappears progressively in the following order: proper names, common nouns, substantives, verbs, interjections and gestures.[22] Bergson classifies amnesias into two categories: in the first category, memory disappears arbitrarily and the forgetting is never final. He, in contrast, regards the second category as the true aphasias, which are explained by Ribot's law.

Bergson says about this law, 'How are we to explain the fact that amnesia here follows a methodical course, beginning with proper nouns and ending with verbs? We could hardly explain it if the verbal images were really deposited in the cells of the cortex' (MM: 151–2). Bergson divides the theory of Ribot into two parts: the dynamic progress of amnesia and the deposition of images in the brain cells. He regards the former highly[23] but rejects the latter.[24]

Bergson bases his view on the argument of Wundt[25] when he points out the inconsistency between Ribot's law and localization theory. It is worth noting that Bergson also relies on the experimental psychologist when he criticizes the negative aspect (the localization of memories) of Ribot's theory.

(4) Bergson rejects the localization of memory-images because of the contradiction between the psychological analysis and the facts of pathology. As we have pointed out, there is some likelihood that Freud had an influence on the discussions in *Matter and Memory*. In addition, Freud quotes Jackson's words in his objections to localization (Freud 1953: 56). We should acknowledge that the theory of Jacksonian dynamics had a wide range of influence on both holism and localization theory, in various ways.

To summarize the preceding: (1) Bergson affirms the dynamic aspect of perception; (2) Ribot regards attention as the movement of arrest, but Bergson considers this explanation inadequate and investigates the positive aspect of attention, invoking experimental psychology; (3) and (4) on the one hand, Bergson dissents from the localization of memories because this theory solidifies the continuous progress of perception and memory into distinct things, but on the other hand, he approves Ribot's law on the dynamic process of amnesia. We can see that Bergson agrees to regard perceptions and memory-images as dynamic processes and refuses to regard them as things.

Thus, Bergson has a strategy to purify the dynamics of memory thoroughly (while accepting the localization of brain functions so long as they are observed facts). In brief, emphasizing the dynamic characteristics of memory, he both approves of Ribot's theoretical framework but also points out its inconsistency in order to undermine its basis.

Hacking stresses that Ribot intends not only to scientize the soul but also to secularize it. According to Ribot's positivistic program, replacing the soul by knowledge about memory should eliminate the metaphysical ego and liberate the soul from religion. Ribot treats the cases of *dédoublement* (dual personality) or of multiple personalities as refutations of the identity of a transcendent ego.[26]

Bergson does not respond to Ribot's argumentation directly, but he alludes to the division of the self and to hypnotism, drawing analogies between them and the phenomena of aphasia in the second chapter of *Matter and Memory*.[27] He suggests that this kind of amnesia[28] is not true amnesia because the lost memories are often recovered.

In an early research paper, 'Unconscious Simulation in the Hypnotic State', Bergson points out the complicated relationship between the hypnotizer and the hypnotized.[29] He suspected that the hypnotized subject was engaged in a particular trick, but he could never bring a subject to confess to it.[30] He guessed that subject was implicitly ordered not to be aware of it.[31] We can find probably similarities between this view of hypnotism and his view on split personalities in the second chapter of *Matter and Memory*. Ribot interprets multiple personalities as a refutation of the unitary self, whereas Bergson does not interpret them as a true division of the self, but a disorder of remembering on the basis of the complicated relationship between the hypnotizer and the hypnotized subject. Both Ribot and Bergson discuss the problem of personality in terms of memories, however. Thus, we can see here that the sciences of memory have already begun to spread broadly in this period.

5 Conclusion

We have attempted to draw a diagram of the scientific background to *Matter and Memory* and to locate Bergson's philosophy on this diagram. In the face of the rise of the sciences of memory, Bergson partly accepts the localization theory of brain function while simultaneously standing for – and thoroughly purifying – a view of memory that sees it as dynamic. If he criticizes aspects of the sciences of memory, it is because a metaphysics (for example, a universal mechanism) that is different from that of the sciences of memory prevents him from developing the dynamics of memory.[32]

So, what comes into being after the refinement of this theory of a dynamic memory? Hacking mentions Bergson and Proust together and says that 'the humanistic reaction was to try an alternative: capture the soul not as science but as narrative' (Hacking 1995: 251). He adds, however, that memory as narrative and the science of memory are branches of the same stem, the secularization of the soul.

We will offer a view that differs from Hacking's. When we develop the sciences of memory, which produce the secularization of the soul, do we also have to eliminate metaphysics? Bergson does not think so. He considers that the old metaphysics, utterly separated from the sciences, will undoubtedly disappear,[33] but he conceives of a new metaphysics verified indirectly by the sciences. Bergson affirms that in this metaphysics the concepts are created,[34] and it is well known that Deleuze inherited and developed this conception in his own way.

Ribot believes that with the secularization of the soul, religions and metaphysics become useless and worthless, whereas Bergson does not think that metaphysics, as a world view guiding the sciences, can be removed from them. Bergson points out, for example, that behind Ribot's positive psychology lies a mechanistic metaphysics (Bergson 1972: 482, 484). When Bergson constructs his system of philosophy on the basis of the sciences of memory, we can affirm that he participates in the secularization of the soul, but he does so in a quite different way from Ribot. He demands that scientists become aware of the metaphysics that unconsciously underlies the sciences. This is not a resurrection of the old metaphysics, however, but the creation of a positive metaphysics. This approach was to be inherited by Maurice Halbwachs (1877–1945), among others (seen by Kurt Danziger in his book *Marking the Mind* as a pioneer of an approach that treated memory on the basis of the interaction between the individual and society rather than confining memory to the individual mind or brain).[35] We can say that Bergson attempted to tame the sciences of memory through his program of positive metaphysics, to borrow an expression from Hacking's *The Taming of Chance*.

Hacking says that we may still be locked into the underlying structure of the sciences of memory created between 1874 and 1886 (Hacking 1995: 197). If he is correct, the diagram of the scientific background of *Matter and Memory* that we present here is related, in some way, with the underlying structure of society and the sciences of today. It would be useful to discuss the possibility of a psychodynamics of memory today, but we have run out of the space. We will deal with this subject later.

This paper is supported by the Grants-in-Aid for Scientific Research: Grant-in-Aid-for Young Scientists B 'Historical study of the influence of physiology and psychology on Bergson's philosophy'.

Notes

1. Hacking divides the sciences of memory into two: surface knowledge and depth knowledge (what Foucault called *connaissance* and *savoir*). The former comprises 'the facts that are discovered in this or that science of memory' (Hacking 1995: 198), and the latter is the knowledge 'that there are facts about memory to be found out' (Hacking 1995). The sciences (a), (b) and (c) belong to the former.
2. Sciences (a), (b) and (c) are sciences of the nineteenth century. To them Hacking adds, from the twentieth century '(d) work at the level of cell biology, transmission across potassium channels and the like' and '(e) computer modeling of memory in artificial

intelligence, parallel distributive processing, and other branches of cognitive science' (Hacking 1995: 199).
3 '[W]e shall find an exaltation of spontaneous memory in most cases where the sensori-motor equilibrium of the nervous system is disturbed; an inhibition, on the contrary, in the normal state of all spontaneous recollections which do not serve to consolidate the present equilibrium; and lastly, in the operation by means of which we acquire the habit-memory, a latent intervention of image-memory' (MM: 98).
4 'It was pointed out again and again by the reagents that mere perception of a number of letters, mere presence together in consciousness, was useless for the purpose of associating and learning; that unless they were able systematically to go through the series at a moderate speed and with a fair amount of attention, no abiding impression was left on the memory' (Smith 1886: 70).
5 '[W]e make use of the fugitive image to construct a stable mechanism which takes its place' (MM: 98).
6 '[T]hese observers proved by experiments that rapid reading is a real work of divination. Our mind notes here and there a few characteristic lines and fills all the intervals with memory-images which, projected on the paper, take the place of the real printed characters and may be mistaken for them' (MM: 126).
7 Rosenfield extracts several paragraphs from Broca's statement of 1861: 'What they have lost is therefore not the faculty of language, not the memory of words, nor the actions of the nerves and muscles needed for the articulation of sounds, but the ... *faculty of coordinating the movements required by articulated language* ...' (Rosenfield 1988: 18).
8 'The phenomena of echolalia will confirm at the following page the existence of the semiautomatic motor accompaniment of speech we hear, preparing Bergson's notion: motor diagram' (Bergson 2012: 361; this is the author's translation from the French).
9 Bergson also mentions Romberg, Voisin, Winslow, Arnaud, Spamer and Sélieux.
10 Sommer refused localization of the centre for reading and he argued against Grashey's position at the annual congress of German psychiatrists in 1891. Bergson invokes Sommer's work at several important points in the second chapter of *Matter and Memory*. Sommer had studied under Wundt, so he had a career both as an experimental psychologist and as a psychiatrist. It is interesting for us to think about his work from the viewpoint of the three sciences of memory.
11 In the following passage, Freud states that in the history of medicine there was a tendency to localize the functions of the mind in the brain (he seems to be thinking of phrenology), and that one might think that science had made remarkable progress, when Wernicke refined localization theory. However, Freud criticizes Wernicke's opinion on the localization of memories. 'Is it justified to immerse a nerve fiber ... with its end in the psyche and to furnish this end with an idea or a memory?' (Freud 1953: 55).
12 Meynert believes nerve fibres connect sensory organs to the cerebral cortices and that there is a one-to-one correspondence between them. By contrast, Freud realizes that a nerve fibre diverges and is connected with other fibres in several strata of the brain, so he supposes that those fibres are organized for diverse functions. The disagreement between them comes from the difference in their anatomical knowledge.
13 'If, moreover, we cast a glance at the minute structure of the nervous system as recent discoveries have revealed it to us, we see everywhere conducting lines, nowhere any centres. Threads placed end to end, of which the extremities probably touch when the current passes: this is all that is seen. And perhaps this is all there is, if it is true that the

body is only a place of meeting and transfer, where stimulations received result in movements accomplished, as we have supposed it to be throughout this work' (MM: 227).
14 On the one hand, psychological analysis (Bain and Ribot) leads us to identify the element of perception with that of memory. '[W]e see that a memory, as it becomes more distinct and more intense, tend to become a perception, though there is no precise moment at which a radical transformation takes place, nor, consequently, a moment when we can say that it moves forward from imaginative elements to sensory elements' (MM: 162). On the other hand, according to the facts of pathology, '[W]e must attribute to perception and to memory separate nervous elements' (MM: 162).
15 'Here again, distinct perception and memory-image are taken in the static condition, as *things* of which the first is supposed to be already complete without the second; whereas we ought to consider the dynamic *progress* by which the one passes into other' (MM: 162).
16 In a variant of *Matter and Memory*, that is 'Mémoire et reconnaissance,' Bergson refers to Jackson's research paper in the footnotes of ninety-two pages.
17 However, *On Aphasia*, the work of Freud mentioned in *Matter and Memory*, belongs to another type of science of memory, that is, the theory of the localization of memories.
18 If we consider the influence of associationistic psychology on Ribot, we can extend this list: Bain, Maudsley, Lehmann, Brochard, Rabier, Pillon, Sully, Höffting, etc.
19 With regard to the idea of psychodynamics, Freud and Ribot are both influenced by Jackson (who, incidentally, is influenced by Spencer). When we take into consideration the spread of his way of thinking, Jackson seems to be the real originator of psychodynamics, but there is another reason (secularization of the soul) why Ribot should be seen as the founder.
20 '... the consciousness of these nascent movements, which follow perception after the manner of reflex, must be here also at the bottom of recognition' (MM: 111).
21 '... we are content to see, in the movements described by Ribot, only the negative condition of the phenomenon. For, even if we suppose that the accompanying movements of voluntary attention are mainly movements of arrest, we still have to explain the accompanying work of the mind, that is to say, the mysterious operation by which the same organ, perceiving in the same surroundings the same object, discovers in it a growing number of things' (MM: 122).
22 'The amnesia affects proper names, which are purely individual, then the names of things which are the most concrete, then all substantives ..., finally the adjectives and the verbs which express qualities, states and acts' (Ribot 1906: 132–3, translated by the author).
23 Ribot insists upon the dynamic characteristics of memory and Bergson approves of this. Ribot cites Meynert's calculation of 600 million cells in the cerebral cortex, but he does not accept the theory that a memory-image is imprinted on a unique cell. He pays more attention to the dynamic associations between the cells than a trace imprinted on each cell, as the basis of memories. 'The organic memory supposes not only a modification of nerve cells, but also the establishment of certain *dynamic associations*' (Ribot 1906: 16, translated by the author).
24 Ribot expresses his localization theory thus: 'In fact, there is not a memory, but memories; there is not a seat of memory, but particular seats for particular memories' (Ribot 1906: 11, translated by the author). It is this aspect of Ribot's view that Bergson rejects.

25 In the discussion to which Bergson refers, Wundt mentions and criticizes Meynert, Munk and Kussmaul (trans. Wundt 1886: 239). Ribot also quotes from Kussmaul's book *Die Störungen der Sprache* when he proposes his law of regression. Kussmaul relates these phenomena to the structure of brain (Kussmaul, 1877: 163–70).
26 Hacking shows that Ribot's program is inherited by Daniel Dennett, author of *Consciousness Explained*. Hacking himself objects to the approach that connects multiple personalities with the metaphysical problem of the relationship between mind and body.
27 Concerning the relationship between multiple personalities and the aphasias, Bergson says, 'Without wishing to be too dogmatic on a question of this kind, we cannot avoid noticing the analogy between these phenomena [of amnesias] and that dividing of self of which instances have been described by Pierre Janet', MM: 151).
28 Bergson classifies amnesias into two types and regards the second type as the true amnesias, which are explained by Ribot's law. He says, 'In the amnesias of the first type, which are nearly always the result of a violent shock, I incline to think that the memories which are apparently destroyed are really present, and not only present but acting' (MM: 150).
29 Ellenberger mentions this rarely discussed paper several times, focusing on Bergson's attitude to hypnotism in his book, *The Discovery of the Unconscious*. When he was a professor of philosophy in Clermont-Ferrand, Bergson participated in hypnotic sessions organized by a physician named Moutin. 'Bergson himself made some remarkable experiments about unconscious simulation of hypnotized subjects' (Ellenberger 1970: 168).
30 One of the experiments is done under the conditions like this: a hypnotizer and a someone to be hypnotized face each other. The hypnotizer opens a book, holding it so the back cover faces the hypnotized, who then divines the page number. Bergson doubts whether hypnotized who becomes hyperesthesia in the hypnotic state reads the very small image of letters (0.1mm) reflected on the cornea of the hypnotizer. He confirms his hypothesis to some extent.
31 'They [the subjects hypnotized] will deny that they had used this means and they will deny sincerely that probably, because they had been ordered implicitly to not become aware of that. . . . Can we say that there is a kind of "unconscious simulation"?' (Bergson 1972: 338)
32 'It is vain, therefore, to treat memory-images and ideas as ready-made things, and then assign to them an abiding place in problematical centers. Nor is it of any avail to disguise the hypothesis under the cover of a language borrowed form anatomy and physiology . . . and since it is born a priori from a kind of metaphysical prepossession, it has neither the advantage of following the movement of consciousness nor that of simplifying the explanation of the facts' (MM: 159–60).
33 Bergson criticizes the old spiritualism thus: 'First of all, I recognize that the upper faculties, that is to say, intelligence, reason, creative imagination, are proper to human beings. . . . However, when the old spiritualism is fought with the materialism and makes an effort to determine the relationship between mind and body, it is entrenched in these upper faculties as in a fortress. In my opinion, this is doubly wrong. It appeared *arbitrary*, and it was *infertile*' (Bergson 1972: 475, translated by the author).
34 'But [metaphysics] is strictly itself only when it goes beyond the concept, or at least when it frees itself from inflexible and ready-made concepts and creates others very different from those we usually handle' (CM: 197).
35 Danziger (2008: 265–6).

References

Bergson, Henri. (1934) *La pensée et le mouvant*. Paris: PUF. (*The Creative Mind*. Trans. M. L. Andison. New York: Philosophical Library. https://archive.org/details/in.ernet.dli.2015.223138/page/n5).

Bergson, H. (1896), *Matière et mémoire*, Paris: Presses universitaires de France (2012) ver. L'édition critique. *Matter and Memory* (1911), trans. N.M. Paul and W.S. Palmer, London: George Allen and Unwin, https://archive.org/details/in.ernet.dli.2015.506437/page/n7

Bergson, H. (1972), *Mélange*, Paris: Presses universitaires de France.

Bernard, D. (1885), *De l'aphasie et de ses diverses formes*, Paris: Progrès médical.

Broadbent, W. (1879), 'A Case of Peculiar Affection of Speech, With Commentary', *Brain: A Journal of Neurology*, 1: 454–503.

Danziger, K. (2008), *Marking the Mind: A History of Memory*, Cambridge: Cambridge University Press.

Ellenberger, H. (1970), *The Discovery of the Unconscious: The History and Evolution of Dynamic Psychiatry*, New York: Basic Books.

Freud, S. (1891), *Zur Auffassung der Aphasien*, Franz Deuticke. *On Aphasia: A Critical Study* (1953), trans. E. Stengel, New York: International Universities Press.

Goldstein, K. (1910), 'Über Aphasie', *Beihefte zur Medizinischen Klinik* 1: 1–32.

Hacking, I. (1995), *Rewriting the Soul: Multiple Personality and the Sciences of Memory*, Princeton: Princeton University Press.

Kussmaul, A. (1877), *Die Störungen der Sprache. Versuch einer Pathologie der Sprache*, Paris: Vogel. *Les troubles de la parole* (1884), trans. A. Rueff, Paris: J.B. Baillière.

Lichtheim, L. (1895), 'On Aphasia', *Brain: A Journal of Neurology*, 7: 433–84.

Moeli, C. (1890), 'Über Aphasie bei Wahrnehmung der Gegenstäde durch das Gesicht', *Berliner klinische Wochenschrift*, 27: 377–81.

Müller, G.E. and F. Schumann (1894), 'Experimentelle Beiträge zur Untersuchung des Gedächtnisses', *Zeitschrift für Psychologie und Physiologie der Sinnesorgane*: 81–190, 257–339.

Ribot, T. (1881), *Les maladies de la mémoire*, 18th edn (1906), Paris: Alcan.

Rosenfield, I. (1988), *The Invention of Memory*, New York: Basic Books.

Skwortzoff, N. (1881). *De la cécité des mots*, Paris: A. Delahaye and E. Lecrosnier.

Smith, W.G. (1895), 'The Relation of Attention to Memory', *Mind: A Quarterly Review of Psychology and Philosophy*, 6, New Series: 47–73.

Wernicke, C. (1874), *Aphasische Symptomencomplex*, Breslau: Max Cohen & Weigert.

Wundt, W. (1874), *Grundzüge der physiologischen Psychologie*, Leipzig: Engelmann. *Eléments de Psychologie physiologique* (1886), trans. E. Rouvier, Paris: Alcan.

Wysmann, J.W.H. (1891), 'Aphasie und verwandte Zustände', *Deutsches Archiv für klinische Medicin*, 41: 27–52.

3

Bergson's Dualism Today

Joël Dolbeault

This chapter aims to discuss a set of arguments for and against psycho-physical dualism as defended by Bergson in *Matter and Memory*. In part 1, I will attempt to explain in what sense Bergson supports a psycho-physical dualism, and in what sense also this dualism differs from Cartesian dualism. In parts 2 and 3, I will discuss some general arguments for and against dualism. In parts 4 and 5, finally, I will discuss some arguments based on scientific ideas. In each part, I will refer to Bergson, but also to some elements of the contemporary debate on mind.

At the time Bergson published *Matter and Memory*, dualism was 'in small honor among philosophers' (MM: xi), and this is still the case today. However, I think that this aspect of Bergsonian thought deserves our interest, on the base of three points. In philosophy today, there is no consensus on the nature of the mind. In neurobiology, the physicalist consensus is limited in scope because neurobiologists do not take into account certain fundamental questions raised by philosophers – in particular the question of whether, from a simple correlation between a mental state and a cerebral state, we can conclude that the two states are identical.[1] In cosmology, finally, we know too little about the universe in general to conclude with certainty that, in the distant past, everything was only matter obeying blind laws.[2]

1 Bergson's position on mind

From the very first page of *Matter and Memory*, Bergson states that his position on mind is 'frankly dualistic' (MM: xi). Moreover, in chapter IV of the book, he says that he wants to correct 'ordinary dualism' (MM: 294, 297), which implies a goal of reconstructing dualism. Therefore, it seems difficult to avoid the conclusion that Bergson supports a kind of psycho-physical dualism. What is interesting is to specify which one.

In *Matter and Memory*, Bergson says that the distinction between mind and matter is profound (MM: 235), radical (MM: 294) and irreducible (MM: 297). Certainly, for him, matter partakes of duration, like mind – a point underlined by several commentators (Jankélévitch 1989: 173–4; Deleuze 1966: 90). Hence, the idea that matter has a kind of memory (MM: 268; ME: 22–3; DS: 47–9). However, between the duration of mind and that of matter, there is *a difference of nature*, not of degree. This difference is mainly due to the fact that matter is incapable of memory par excellence. It cannot retrieve its past in the form of representations, only to repeat it in the form of actions (MM: 297–8).[3]

Moreover, Bergson considers that the mind could exist without physical support (CE: 283; ME: 35–6, 71–4; CM: 52–3; DSMR: 251–4).[4] In *Matter and Memory*, this corresponds to the idea that 'memory is something other than a function of the brain' (MM: 315, 317), and that pure memories are preserved without physical support (MM: 191–3, 231–2). Bergson also writes that memory is 'a power absolutely independent of matter' (MM: 81, 232). In the terms of contemporary philosophy of mind, this corresponds to the idea that memory is not a *property* of the brain, but a kind of substance, in the sense of a thing capable of subsisting by itself. Conclusion, Bergson supports a kind of dualism of substances, as opposed to the dualism of properties.

Now, Bergson's dualism is very different from Descartes's dualism (which is the reference dualism in the debate on mind), and this difference concerns both matter and mind.

For Descartes, matter has only one quality: extension (Descartes 1985: 210–11, 224). Moreover, material changes are all reduced to movements (Descartes 1985). Finally, matter is devoid of any psychic property (Descartes 1985). In contrast, for Bergson, matter is a set of diverse qualities, some of which are at the origin of the qualities we perceive: colours, sounds, smells, etc. (MM: xii–xiii, 26–7, 238, 291–2). Moreover, matter is intrinsically temporal in the sense that it is a succession of elementary vibrations (MM: 267–77). This amounts to saying that every material thing is reduced to a continuity of events (CM: 84–5, 172–3). Finally, matter has psychic properties, a motor memory in particular. I come back to this idea in parts 2 and 3.

Let us now turn to the mind. For Descartes, the mind is not extended (Descartes 1985: 211). Moreover, Descartes is not interested in the temporality of the human mind in the sense that, for him, time (physical or psychic) is only a succession of distinct states (Descartes 1984: 254–5).[5] By opposition, for Bergson, the mind has degrees of extension according to its activity. Thus, in perception, it extends and coincides partially with matter (MM: 288–94). Moreover, in addressing the temporality of the human mind, Bergson makes the hypothesis that each of our perceptions contracts many elementary material vibrations (MM: 267–77), and that something equivalent exists for our actions: a contraction of the infinitesimal indeterminations of matter (ME: 20–2).

From this new conception of matter and mind, a new conception of psycho-physical interaction follows. In classical dualism, this interaction remains something obscure (MM: 235–6, 294–5). In contrast, in the dualism proposed by Bergson, this interaction becomes clearer through the introduction of two fundamental ideas. The first idea is that *the mind has degrees of extension*. Thus, in perception, it extends to coincide with matter; in action, it also extends to a certain point since, in Bergson's own words, to go from the general scheme of the action to its sensory-motor components is to go 'from the intensive to the extensive, from a reciprocal implication of parts to their juxtaposition' (ME: 230). The second idea is that *the mind is capable of contracting micro-events*, both in perception and in action. In action, this implies that it can act on the body (the brain in particular) without having to represent the atomic or molecular detail of its action (ME: 20–2).

2 General arguments in favour of interactionist dualism

Dualism is 'suggested by the immediate verdict of consciousness' (MM: xi). In other words, it is consistent with appearance: the appearance of a difference in nature between mind and matter, but also the appearance of an interaction between them (MM: 4–5; ME: 39). Now, methodologically, Bergson defends the idea that what appears clearly to consciousness must be considered as real, unless one proves that it is only an appearance.[6] From this point of view, for him, the burden of proof falls on the theories that contest dualism, that is, the appearance of things.

A first contestation may concern the idea of a difference in nature between mind and matter. This is the object of the theories of psycho-cerebral identity in philosophy of mind. The problem is that these theories are directly opposed to what appears: mental states are not identical to cerebral states. Some philosophers have thought that, to get out of the difficulty, it was possible to defend the idea that mental states are reduced to functions. Again, however, the problem is that this idea is directly opposed to what appears: at least some mental states, if not all, are characterized by *qualia*. Consequently, the various theories of psycho-cerebral identity have reached an impasse.

A second contestation may concern the idea of an interaction between mind and matter. This is the object of the epiphenomenalist theory, very widespread at the time of Bergson (MM: xiv–xv, 15–16). This theory does not contest the irreducibility of the mental to the cerebral. It only contests the idea that the mental could act on the cerebral. In this sense, for epiphenomenalism, the cerebral life determines the mental life, but never the opposite. The problem is that this theory is also opposed to what appears to consciousness, namely that our mental life determines some of our actions. Another problem is that this theory tends to contradict science. According to biology, indeed, the enduring characteristics of living species have practical utility because these characteristics have been selected for by evolution. Since mental states are an enduring characteristic of humans (and probably of pre-humans), this implies that they must have practical utility, i.e. a practical effect. This last argument, developed by James (1879) and taken up by Popper (1977: 72–5), has largely contributed to the abandonment of epiphenomenalism by philosophers.

Another way of rejecting dualism may be to accept the idea of a certain difference in nature between mind and matter, but to attenuate this difference by affirming that mind is not substantial, in the sense that it would emerge from matter (according to certain physical conditions). Compared to the previous theories, this approach has the merit of being compatible with what appears to consciousness. However, this approach is not logically opposed to interactionist dualism, since it is compatible with the idea that the (emergent) mind is distinct from matter and that it interacts with it – a position that Popper (1977: 22–35) defends. On the other hand, the idea that something radically different from matter can emerge from matter is problematic for many philosophers. This explains the interest in panpsychism today in philosophy of mind.

The fundamental intuition of contemporary panpsychism is that elementary matter possesses certain mental properties, which makes it unnecessary to conceive of the emergence of the mental from the physical. From this point of view, panpsychism is a response to emergentism. However, is it also a response to dualism? To say that, fundamentally, reality is made up of mental and physical properties, is it not to generalize dualism to the whole of nature? Of course, contemporary panpsychism takes many forms today which it is not possible to discuss here. However, whatever its form, if panpsychism wants to save *what appears to consciousness*, is it not obliged to affirm the existence of two different and interacting realities?

Bergson also defends a form of panpsychism. However, for him, panpsychism does not answer the question of the nature of the mind. In other words, it does not aim at replacing psycho-physical dualism. For Bergson, panpsychism answers the question of causation within matter. His main idea is that, to explain physical regularities, it is necessary to postulate a kind of motor memory immanent to matter. From this follows the idea that matter is capable of perception (MM: 30-1, 46), of memory of the immediate past (ME: 22-3; DS: 47-9) and of motor reaction (MM: 297-8). Matter is what Bergson calls a 'neutralized consciousness' (MM: 313, 331).[7]

3 General arguments against interactionist dualism, and replies

From Descartes until today, there have been many arguments against interactionism. However, despite this diversity, most arguments are built on the same pattern:

(i) First, a fundamental principle is admitted about causation. For example:

- Two realities A and B, the first one being spatial, the second one being not spatial, cannot interact.[8]
- Two realities A and B which are radically different cannot interact.[9]
- Any action of a reality A upon a reality B must respect the principle of conservation of energy.[10]
- Any action of a reality A upon a reality B must respect the laws of physics (principle of causal closure of the physical domain).[11]

(ii) Secondly, it is claimed that interactionist dualism does not respect this fundamental principle.
(iii) Finally, it is concluded that interactionist dualism cannot be true.

There are several replies to this kind of argument, but the main reply is to criticize the principles laid down at the start. These principles are not *a priori* truths. They can only have an empirical value. Consequently, they are only broad extrapolations, and we must not mistake these extrapolations with necessary truths. This reply can be found in Bergson (ES: 43-5), but also in Popper (1977: 176-82). According to Popper, this kind of argument against dualism is based on an essentialist conception of causation, as opposed to a conjectural conception. The essentialist conception claims to know the ultimate truth about causation, but this approach to reality is opposed to the spirit of science.

Some might reply that the first two principles have an *a priori* value, because they deal with what would be *inconceivable* in terms of causation. However, since Hume, we know that this position is untenable. Hume shows indeed that a causal assertion between two facts A and B does not require the spatiality of these facts, nor their similarity. This assertion relies only on the constant conjunction between A and B. This is why Hume writes that 'anything may be the cause or effect of anything' (Hume 2007: 163), specifically about the mind-body relation.

Now, let us imagine that we want to go beyond Hume by developing a theory of causation that is more explanatory than his, in the sense of a theory that aims at explaining physical regularities (and not only at noticing them). Metaphysics then offers the choice between three theories. The first is that physical regularities are due to *laws* capable of governing the behaviour of matter. The second is that these regularities are due to *dispositions* immanent to matter, and capable of governing its behaviour. The third (defended by Bergson, Peirce and James) is that these regularities are due to a kind of *motor memory* immanent to matter, and capable of governing its behaviour (Dolbeault 2021). In the three cases, however, the principles stated above against dualism lose all value; indeed, one considers that a non-spatial reality (laws, dispositions, or memory) acts on something spatial. And in all three cases, one considers that this non-spatial reality, far from being subject to natural regularities, is at the origin of them. Consequently, the deeper one goes into the metaphysical reflection on causation, the more arbitrary the principles stated against interactionism seem.

4 Libet's experiment

In the contemporary debate on the mind, there is a well-known experiment which seems to embarrass the dualistic position: the experiment conducted in the 1980s by Benjamin Libet on voluntary action. Let us recall that the aim of this experiment is to determine if the conscious will to act is temporally prior or posterior (possibly simultaneous) to the brain activity that leads to the action. If it is prior, we should infer that voluntary action is engaged and specified by consciousness. If it is posterior, we should infer that it is engaged and specified by brain activity – consciousness then appearing as a mere epiphenomenon (Libet 2004: 123).

In short, the protocol of the experiment is the following one: you ask a subject to look at a clock suitable to the experiment, and to pay attention to the time when he or she spontaneously decides to perform an action. The action is to flex the wrist. The subjects repeat the process several times, and each time, their brain activity is measured by electroencephalography. The result of the experiment is that the moment when the subjects relate their decision to act is subsequent to the activation in the Supplementary Motor Area (SMA) which engages the action. More specifically, for activation in the SMA at -550 msec,[12] this moment only happens at -200 msec, and the activation of the

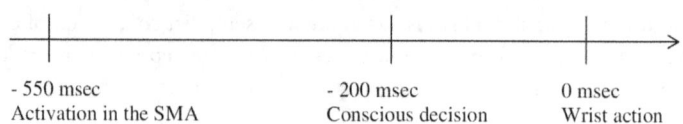

Figure 3.1 Libet's experiment, results. © Joël Dolbeault.

muscle at 0 msec (these measures are obviously averages). For some, this experiment would therefore prove that the conscious will to act is produced and determined by brain activity, and psycho-physical dualism would be false.

However, many thinkers reject this conclusion.[13] Their criticisms differ in detail, but we always find the same idea, namely that Libet's experiment is not about voluntary action, *because of the protocol of the experiment*. The protocol says that the subjects should not plan their action; they should wait until they feel a spontaneous urge to move (Libet 2004: 126). However, waiting for a spontaneous urge to move and deciding something consciously are two very different things. In the first case, it is a passive attitude. In the second case, it is an active one, with reasoning, deliberation, etc. Thus, the experiment is supposed to focus on the conscious decision to act, but it paradoxically places the subjects in a situation of unconsciousness.

In fact, the conscious decision of the subjects takes place before the experiment, when they agree to participate and to respect the protocol, but during the experiment, they are only waiting for a spontaneous urge to act, the cause of which is then unconscious. It is therefore normal that the brain activation that gives rise to this urge precedes the consciousness of this urge.

Other experiments were conducted following Libet, which largely confirm his results (Soon et al. 2008). However, each time the protocol is similar; the spontaneous urge to move, without reason, is taken as a model for conscious decision. Therefore, each time the same criticism can be made.

From a Bergsonian point of view, this criticism is valid. Let us recall that, for Bergson, consciousness means memory and anticipation (ME: 7–8). Therefore, a conscious action is an action preceded by reflection. By reducing memory and anticipation to zero in the protocol, the Libet-type experiments do not deal with conscious action at all.[14]

5 Dualism and memory

On memory, Bergson's dualistic position can be summarized as follows: the brain contains the mechanisms that allow memories to be retrieved, but it does not contain the memories themselves. What should we make of this hypothesis in the light of current neuroscience? A serious answer to this question requires a long and difficult work that remains to be done. However, some basic remarks are necessary, in particular to answer the objections made to Bergson.[15]

First, it is obvious that Bergson's hypothesis predicts the identification of neural correlates of memory, at several levels (level of the general organization of the brain, molecular level, etc.). The identification of such correlates is not at all an objection against Bergson. The question is whether these correlates are used *to store* memories, or only *to retrieve* them.

Second, it is also clear that Bergson's hypothesis predicts a causal link between brain damage and amnesia. The question is how to interpret this link: as the destruction of certain memories or only as the destruction of what allows memories to be retrieved.

Third, the hypothesis that memories are stored in the brain encounters a difficulty. The storage of a memory is supposed to correspond to a vast network of interconnected neurons in distinct areas of the brain, areas that are specialized in the processing of shapes, colours, movements, and so on. However, the fact that these multiple areas can correspond to a single representation is not well understood. In the literature, we speak of 'the binding problem'. The general opinion is that this problem is not solved.[16] It concerns memories, perceptions and more generally all forms of conscious representation.

Fourth, an argument developed by Tulving (a psychologist specializing in memory) seems to go in the direction of Bergson. Tulving points out that it seems impossible to identify a mnemonic engram independently of the act of retrieval. A mnemonic engram is a large network of interconnected neurons, but this network can only be observed during the retrieval process. His hypothesis is therefore as follows:

> However, if the engram is a kind of an entity that manifests itself only in activity, or retrieval, then we might conjecture that the physical changes resulting from an experience *do not exist as an engram in the absence of that activity*.
>
> Tulving 1991: 91

> Of course, memories are not stored at synapses, but I think it is useful to contemplate the possibility that they are *not stored anywhere else in the brain either*. The whole issue of where or, more important, how memories are stored in the brain may turn out to be an incorrect formulation of the problem, despite its seductively enticing allure. And the source of such incorrect formulation may lie in the single-minded preoccupation with the storage, or the engram, and sometimes even *identification* of storage with memory.
>
> Tulving 1991: 90–1

In the first passage, Tulving emphasizes the fact that the brain engram that is supposed to correspond to the storage of the memory has only an intermittent existence in the brain: when the memory is retrieved, the engram appears physically; when the memory is not retrieved, the engram does not appear.

In the second passage, he clearly makes the hypothesis that this engram has no lasting existence, and that it corresponds only to a certain activity of the brain. In other words, he makes the hypothesis that memories are not stored in the brain. Of course, this is only a hypothesis, and few neuroscientists follow Tulving on this point, but, as I said in the introduction, it also appears that few neuroscientists are sensitive to the fundamental questions raised by philosophers.

Conclusion

Bergson's dualism differs from Cartesian dualism on many points: on matter, on mind, and on the interaction between the two. From this point of view, it is an almost entirely new theory.

According to Bergson, this theory is suggested by the immediate data of consciousness: by the obvious difference between the external world and the internal life, and by the obvious interaction between the two. Therefore, the burden of proof is on the non-dualistic theories of mind. While the various theories of identity stumble over *qualia*, contemporary panpsychism tends to assert the existence of a fundamental duality, in the whole of nature, between physical and mental properties.

Some principles have been formulated against dualism, but these principles seem arbitrary: they are based on an essentialist conception of causation, without solid justification. On the other hand, the more empirical challenge to dualism, based on Libet's experiments, raises major difficulties.

Bergson's psycho-physical dualism still deserves our interest today.

Notes

1. On this weakness of theories of consciousness in neurobiology, see Niikawa (2020: 8–10).
2. Some physicists consider that, faced with a certain number of enigmas (dark matter, dark energy, fine tuning, etc.), cosmology is in crisis. See in particular: Smolin (2006), Hossenfelder (2019).
3. Let us also specify that, contrary to what Deleuze claims, Bergson never spoke of a kind of undifferentiated duration that would differentiate itself into matter and life (Deleuze 1966: 104–6). For Bergson, a supra-consciousness created matter and life, and this creation is not at all a differentiation (CE: 260-2).
4. This statement is not the same as saying that the mind is immortal, because this point is not empirically provable, since all experience concerns a limited time (MM: 72; CM: 52-3).
5. In his lectures at the Collège de France, Bergson addresses this criticism to Descartes (Bergson 2016: 292–4).
6. For this principle of method in Bergson, see in particular (DS: 38).
7. On Bergson's panpsychism and its relation to contemporary panpsychism, see Dolbeault (2018).
8. See in particular Kim (2005: 70–92).
9. See in particular: Ryle (1949: 12–13), Churchland (1986: 318–20).
10. See in particular: Bunge (1980: 17), Dennett (1991: 33–5).
11. See in particular: Papineau (1993: 16–17, 29–32), Kim (2005: 15–16).
12. Milliseconds.
13. See notably: Batthyany (2009), Nahmias (2010), Mele (2011).
14. For more on a Bergsonian critique of Libet's experiments, see Dolbeault (forthcoming).
15. See in particular: Missa (1993: 83–4, 137–63), Delacour (1997), Beaunieux, Desgrange and Eustache (2000).
16. See in particular: Kandel (2007: ch. 22), Deheane (2014: 191–192).

References

Batthyany, A. (2009), 'Mental Causation After Libet and Soon: Reclaiming Conscious Agency', in A. Batthyany and A. Elitzur (eds), *Irreducibly Conscious*, 135–61, Wien: Universitätsverlag Winter Heidelberg.

Beaunieux, H., B. Desgrange and F. Eustache (2000), '*Matière et Mémoire* et les Modèles Actuels de la Mémoire: de Bergson à Tulving', in B. Claverie, R. Jaffard, B. Andrieu (eds), *Bergson: de la Philosophie aux Neurosciences*, 95–112, Bordeaux: Université Victor-Segalen.

Bergson, H. (1911), *Matter and Memory*, trans. N.M. Paul and W.S. Palmer, London: George Allen and Unwin, https://archive.org/details/in.ernet.dli.2015.506437/page/n7

Bergson, H. (1911), *Creative Evolution*, trans. A. Mitchell, London: Macmillan, https://archive.org/details/in.ernet.dli.2015.94170/page/n5

Bergson, H. (1920), *Mind-Energy*, trans. H. Wildon Carr, New York: Henry Holt and Co, https://archive.org/details/mindenergylectur00berguoft/page/n6

Bergson, H. (1965), *Duration and Simultaneity*, trans L. Jacobson, Indianapolis: Bobbs Merrill, https://archive.org/details/DurationAndSimultaneityHenriBergson/page/n3

Bergson, H. (1946), *The Creative Mind*, trans. M.L. Andison, New York: Philosophical Library, https://archive.org/details/in.ernet.dli.2015.223138/page/n5

Bergson, H. (1935), *The Two Sources of Morality and Religion*, trans. R. Ashley Audra and C. Brereton, New York: Henry Holt, https://archive.org/details/in.ernet.dli.2015.191246

Bergson, H. (2016), *Histoire de l'Idée de Temps*, Paris: Presses universitaires de France.

Bunge, M. (1980), *The Mind-Body Problem*, Toronto: Pergamon Press.

Churchland, P. (1986), *Neurophilosophy*, Cambridge, MA: MIT Press.

Dehaene, S. (2014), *Le Code de la conscience*, Paris: Odile Jacob.

Delacour, J. (1997), '*Matière et mémoire* à la lumière des neurosciences contemporaines', in P. Galois and G. Forzy (eds), *Bergson et les neurosciences*, 23–7, Le Plessis Robinson: Institut Synthélabo.

Deleuze, G. (1966), *Le bergsonisme*, Paris: Presses Universitaires de France.

Dennett, D. (1991), *Consciousness Explained*, New York: Little, Brown & Company.

Descartes, R. (1984), 'Principles of Philosophy', in *The Philosophical Writings of Descartes*, vol. 2, Cambridge: Cambridge University Press.

Descartes, R. (1985), 'Principles of Philosophy', in *The Philosophical Writings of Descartes*, vol. 1, Cambridge: Cambridge University Press.

Dolbeault, J. (2018), 'Bergson's Panpsychism', *Continental Philosophy Review*, 51 (4): 549–64.

Dolbeault, J. (2021), 'Laws, Dispositions, Memory: Three Hypotheses on the Order of the World', *Metaphysica*, 22 (1): 101–21.

Dolbeault, J. (forthcoming), 'La libre volonté selon Bergson: Éclaircissements conceptuels et confrontation aux expériences de type Libet', *Intellectica*.

Hossenfelder, S. (2019), *Lost in Math: How Beauty Leads Physics Astray*, New York: Basic Books.

Hume, D. (2007), *A Treatise of Human Nature*, Oxford: Clarendon Press.

James, W. (1879), 'Are We Automata?' *Mind*, 4 (13): 1–22.

Jankélévitch, V. (1959), *Henri Bergson*, Paris: Presses Universitaires de France.

Kandel, E. (2014), *In Search of Memory*, New York: W.W. Norton.

Kim, J. (2005), *Physicalism, or Something Near Enough*, Princeton: Princeton University Press.

Libet, B. (2004), *Mind Time: The Temporal Factor in Consciousness*, Cambridge: Harvard University Press.

Mele, A. (2011), 'Free Will and Science', in R. Kane (ed.), *The Oxford Handbook of Free Will*, 2nd edn., 499–514, New York: Oxford University Press.
Missa, J.-N. (1993), *L'Esprit-Cerveau*, Paris: Vrin.
Nahmias, E. (2010), 'Scientific Challenges to Free Will', in C. O'Connor and T. Sandis (eds), *A Companion to Philosophy of Action*, 345–56, Malden: Wiley-Blackwell.
Niikawa, T. (2020), 'A Map of Consciousness Studies: Questions and Approaches', *Frontiers in Psychology*, 11: 1–12, https://www.frontiersin.org/articles/10.3389/fpsyg.2020.530152/full
Papineau, D. (1993), *Philosophical Naturalism*, Oxford: Blackwell.
Popper, K. and J. Eccles (1977), *The Self and Its Brain*, Berlin: Springer.
Ryle, G. (1949), *The Concept of Mind*, London: Hutchinson's University Library.
Smolin, L. (2006), *The Trouble with Physics*, Boston: Houghton Mifflin Harcourt.
Soon, C.S. et al. (2008), 'Unconscious Determinants of Free Decisions in the Human Brain', *Nature Neuroscience*, 11 (5): 543–5.
Tulving, E. (1991), 'Interview with Endel Tulving', *Journal of Cognitive Neuroscience*, 3 (1): 9–94.

4

Memory and History: Rereading Bergson through Ricœur

Masato Goda
Meiji University

1

'The remembrance of forbidden fruit [*le souvenir du fruit défendu*] is the earliest thing in the memory of each of us, as it is in that of mankind.' This is the opening sentence of the *Two Sources of the Moral and the Religion* by Bergson (TS: 1). Although it includes the two keywords, *souvenir* and *mémoire*, little study has been done to explore the tie which should bind *Matter and Memory* with *Two Sources*, at least as far as I know. How could memory of each of us relate itself with memory of the humankind? How could individual memory interweave with collective history or evolution of mankind? An immense question which I cannot, of course, fully answer, but today I would like to venture toward this problematic.

For this purpose, I'd like to draw a kind of an auxiliary line. That is, I would like to present you with an extract from a dialogue between a scientist and a philosopher. First, the scientist says: 'Bergson affirmed that in principle memory has to be a power which is absolutely independent from matter and that the attempt to derive remembrance from the brain would have to appear as fundamental illusion for the analysis'. With regard to this topic, the intuition held by this great philosopher turns out to be wrong ... The neuronal inscription of the trace is thus obvious. But even so, we still must make much effort in order to decipher the synaptic hieroglyphs.'

In responding to such a severe assessment of Bergson, our philosopher didn't reply directly. Rather, he shifted the topic by referring to the distinction Bergson made between memory-habitude and memory-remembrance and, as for the latter, he mentioned the notion of 'dynamic schema' associated with the names of Bergson and of Merleau-Ponty, in order to affirm the 'evident coincidence between neuro-biological approach and phenomenological thesis of this notion of dynamic schema'.

Who are the scientist and the philosopher? The scientist is Jean-Pierre Changeux, author of *Neuronal Man: The Biology of the Mind* (1985), and the philosopher is Paul Ricœur. Their dialogue took place in 1999. I don't know whether it was a necessary encounter as they themselves acknowledged or a rather unfortunate '*misencouter*' as

Catherine Malabou notes in her *Que faire de notre cerveau?* (2011). Suffice to say that, by suspending the negative judgement by Changeux on Bergson, Ricœur seemed to situate his own task within the perspectives opened by Bergson's *Matter and Memory*. You may think I am perhaps exaggerating the link between Bergson and Ricœur; nevertheless, not only the very dialogue with Changeux but also the title of his book *Memory, History, Forgetting* (2006) seems to confirm the strength of this link. In addition, one of the disciples of Ricœur, Jean-Luc Petit, testifies to this link indirectly through his own researches that associate Husserl's phenomenology with the data gathered by contemporary neurosciences.

Today, I'll try to read Ricœur's *La philosophie de la volonté* in the light of Bergson's works in the first section. This will be followed by my re-reading of *Matter and Memory*, a reading which will reflect vis-à-vis my first section on Ricœur.

2

Ricœur's doctoral thesis *The Voluntary and the Involuntary* (*Le volontaire et l'involontaire*) was published in 1950. In its general introduction, Ricœur cites the following passage from a letter of Descartes addressed to Elisabeth, suggesting that his main concern lies precisely in this mind-body problem just as for Bergson in *Matter and Memory*: 'the things belonging to the union of body and soul are clearly understood by the senses' (Ricœur 2009: 26). The philosophy of will conceived by Ricœur aims to explore the 'total field of motivations' (*champ total de motivations*) of our action or project in its large sense by plunging into the simultaneously Cartesian *and* Husserlian Cogito's underground; Ricœur uses the word '*Abgrund*' (Ricœur 2009: 104) to designate its darkness (*ténèbres*). In this respect, Ricœur, in following the way paved by Maine de Biran, tries to think how the habit, the unconscious, the corporality function involuntarily.

However, what is to be noted here is the fact that, in addition to these, Ricœur takes account of the 'historicity' (*historicité*) as evident in the title of a section of his book: 'The plan of history and the plan of body' (*Le plan de l'histoire et le plan du corps*). In contrast, we cannot find the word 'history' as a central key term in *Matter and Memory*. I may also add that Walter Benjamin pointed out this absence in his 'On Some Motifs of Baudelaire'. To contrast even further – in fact, I don't know whether it is a contrast or not – Ricœur wrote 'Husserl et le sens de l'histoire' in 1949 and asked how Husserl's phenomenology could or couldn't tackle the problematic of history in its last period. I think Ricœur was very conscious of the fact that Heidegger's *Sein und Zeit* is cut short with the chapters on temporality and historicity. It is worthy to remark that Ricœur used in the article cited above the expressions such as 'exegesis'(*exégèse*) and 'cultural character of history' (*caractère historique de l'histoire*). It shows that Ricœur already planned for 'the graft [*greffe*] of phenomenology on hermeneutics' in his early work.

Perhaps I've advanced my argument rather hastily. Let us return to the 'union of body and soul', which Ricœur presented by citing the letter of Descartes. In regard to this, Ricœur isn't monist nor dualist. I'd like to cite a remarkable passage from Ricœur:

> Why is the dualism of soul and body a doctrine of entendement (Verstand)? Why is the dualism, in its virtual form of liberty and necessity, almost invincible? There is no reason except this: the tear (déchirure) in this union isn't only an infirmity of Verstand unable to subsume the mystery of union of soul and body, but to some extent a wound or lesion (lésion) of being itself. It isn't only by thinking to this wound that we break the living union of human being; it is precisely in the human act of existing that the secrete tear is inscribed'.
>
> Ricœur 2009: 555

This is why Cogito is possible only insofar as it is shattered. To use one of Ricœur's key terms, Cogito is always 'Cogito *brisé*'. Here I venture to advance my first hypothesis that it is exactly in this '*secrete rift*' within the union of soul and body that Ricœur tries to find the very union and names the union or link the '*polemical* link'. Ricœur emphasizes the adjective 'polemical' by italicizing it, probably in order to throw into relief the etymological sense of this term, that is, war (*polemos*). This reminds me of an anecdote that, at the end of his life, Ricœur was very impressed with a remark of Derrida: 'Je suis en guerre avec moi-même' (I'm at war with myself).

From this point of view, the first to be noticed is that the above-mentioned 'tear' or 'rift' represents the original sense of 'symbol'. Needless to say, this is the notion that traverses the entire career of Ricœur. In the second volume of *The Philosophy of Will*, we find a very interesting passage concerning this notion.

> Cosmos and Psyche are two poles of a same expressivity [of symbol].... This double expressivity has its complement in the third modality of symbol, that is poetical imagination.
>
> Ricœur 1960: 20

Isn't this a passage with a Bachelardian tone? That's how symbol comes to be associated with image in the crack of a human existence; *Abgrund*, darkness, image – these words send me back to the *Einbildungskraft* (force of imagination) as presented by Kant. I've always been convinced that Kant's theory of imagination constitutes the background for Ricœur's philosophy of will as well as for Bergson's *Matter and Memory*. Kant defined it as 'common and unknown radix' from which bifurcate both sensibility and understanding. The imagination, either productive or reproductive, functions as schematism, one which Kant compared to the 'hidden art in the darkness of the soul'.

According to Ricœur, the union of soul and body constitutes itself in the very rift of this union; he names this dynamic process 'becoming-human body' (*devenir-corps humain*) by habit in its largest sense. You may recall here that Bergson spoke about the '*turning point* of experience from which experiences must become *human* experiences' in *Matter and Memory* (Bergson 1959: 231). What the word 'body' designates here must be an 'image- or scheme-body' and 'symbol-body', from which bifurcate two poles of the so-called soul and the so-called body, the so-called physiological and the so-called psychological. This is the second point I'd like to emphasize today. Bergson himself called this rift or in-between 'region of images' (Bergson 1959: 273), which, according to the expression used in *Matter and Memory* '*existe rather virtuellement*' (Bergson

1959: 371). At this point, what should be underlined is that what we call brain or neuron-system is originally nothing but a provisional coagulation of dynamically vibrating process of image or scheme.

It is very possible to relate what we call 'image-or scheme-body' with '*Körperbild*' or '*Körperschema*' discussed by Paul Schilder, for example. Ricœur himself invokes Merleau-Ponty, Kurt Goldstein as well as Viktor von Weizsäcker in order to reject purely mechanical '*Reflexgesetz*' and to replace it with the 'know-how preformed before any training' (*savoir-faire preformé avant tout apprentissage*) which can serve as the bottomless bottom (*fond*) for actions.

Habit, of course, has many levels. The 'image-body' and 'symbol-body' can be formed by more or less voluntary actions or trainings that often become unconscious or forgotten little by little. In contrast, it is possible that 'image-body' and 'symbol-body' have been formed only through more or less involuntary processes. However, in the extreme, we can't recall them, no matter what efforts we make; in other words, a 'remembrance' (*souvenir*) must be endless in regards to them. With respect to this point, Ricœur writes as follows.

> The involuntary which remain relative to the possible will drives us to the neighborhood of the absolute involuntary; the relative darkness announces the extreme darkness which would be the hidden (*caché*)'.
>
> Ricœur 1960: 474

> The reflection on the habit, on this familiar but enigmatic power, is a kind of invitation to the remembrance without end which loses in the darkness. Then the shift has made from the abolished to the forbidden.
>
> Ricœur 1960: 476

Here Ricœur joins the first phrase of Bergson's *The Two Sources*; I think he is conscious of this crossing or coincidence. In fact, he mentions Bergson's last work, which touches upon the notion of 'obligation.' And I would like to compare Ricœur's passage from the remembrance to the absolutely involuntary, to the hidden and forbidden darkness to that from '*souvenir-image*' to '*souvenir pur*' or '*mémoire pure*' formulated in *Matter and Memory*. However, I should immediately add that, unlike *Matter and Memory* at least on the surface level, Ricœur's research is concerned not only with 'memory' but also with 'history'. He writes explicitly that 'history and memory are two plans of the motivation, two radixes of the involuntary' (Ricœur 1960: 165), and that the motifs not only 'figure' but also 'historicize'. The verb 'figure' could be interpreted as corresponding to the 'formation of image' and '*Ein-bildung*'. But what does 'historicize' mean? What is 'history'? How do 'memory' and 'history' intertwine?

Let us now return to the 'tear' of human existence, where 'body' in the sense of image or symbol forms itself. Following Gabriel Marcel, Ricœur calls it 'mystery of incarnation'; for him 'mystery of incarnation' is nothing but that of memory and history. That is why Ricœur says that 'history leans me as if it were body'. History as well as memory concerns both the voluntary and the involuntary. To give an example, when I see a tree, memory, voluntary or otherwise – in other words, image, scheme, category in the

Kantian sense – infiltrates this perception. Without it, I cannot recognize a tree as a tree. That leads me to presume that language intervenes in one way or another during such process. Image, symbol, category, language – despite their subjective and individual nuances – must be collective and historical products. Marx said that the elaboration of our five senses is a result of world history. In this sense, all behaviour, including what Husserl called 'passive synthesis', is historical, but this does not deny at all yet another assertion that only the temporal being is historical, as Heidegger said in the final part of *Sein und Zeit*.

I have until now regarded the two terms, image and symbol, as synonyms. I think it is impossible to distinguish them substantially; nevertheless, *grosso modo*, symbol seems to be ordinarily seen as more historical and more collective than image. An image as well as a symbol supposes very often some story (history); nevertheless, the story a symbol posits seems to or may be usually more collective than one that an image posits. I am well aware that these remarks are far from being satisfactory, but what I would like to keep in mind about them is that Ricœur has opened the perspective in which we can associate '*souvenir-image*', '*souvenir pur*' in *Matter and Memory* with the notions of 'fabulation' and 'fabulous function' in *Two Sources*. This is the third hypothesis, which we have to consider next. Before closing this part, I should remark, last but not least, that Ricœur, just like Bergson in *Matter and Memory*, employs the phrase 'vibration' of the body to explain the penetration of the social into the individual. Please do keep in mind here what I have said above about the 'in-between'.

3

'Fabulation' is compared by Bergson to 'virtual instinct'; it constitutes the opposite side of pure intelligence and functions not only as sources of diverse myths or mythology, but also as the basis on which all literature will be constructed. All this in order to defend against the dissoluble force of pure intelligence and to recover a tie between society and individuals. So, it plays an essential function proper for the closed society as well as for its static religion and morality. We get to know the products of fabulous function by means of reading, listening, seeing, touching, feeling etc. Bergson writes:

> The static religion attaches a human being to life, therefore the individual to the society by telling him the stories (histoires) like the ones by means of which we lull the baby to sleep. These are not certainly simple stories. Produced necessarily by fabulous function, not for simple pleasure, these stories counterfeit the perceived reality to such an extent that they could prolong themselves into actions.
>
> Bergson 1959: 1154

This passage shows clearly the influences of fabulated stories upon the perception-action; if it were the case, it is natural that they also produce effects upon '*souvenir-image*'. On this, Ricœur develops a very rich discussion in *Memory, History, Forgetting* by examining one by one the cases of Heidegger, Maurice Halbwachs, Yosef Haïm Yerushalmi, Pierre Nora and so on. Being unable to take up this part of his reflections today, I limit myself to remark that, in this text, Ricœur has come to emphasize the role of 'forgetting' as conditions sine qua non of memory and that, from this point of view, the 'pure memory', just like the hidden and absolute involuntary in Ricœur, turns out to be a kind of 'forgetting'; for Ricœur, the field of the absolute involuntary is filled with the symbols of mythology.

As I have suggested above, Walter Benjamin wrote in his 'On Some Motifs of Baudelaire' that 'the Bergsonian notion of duration is separated from the category of history (as well as from that of prehistory) precisely because it lacks the notion of death. Correlatively, it lacks the concept of action'. Benjamin was not wrong, but a passage in Bergson from 'involuntary memory' to 'history' has just opened or reopened by tracing an auxiliary line of Ricœur. This is the fourth point that I want to insist upon.

'Pure memory' is a memory without an image. For example, when did I recognize my face in a mirror for the first time? From when did I have an image of myself? I don't know, I can't know. We have a period of life, which can't be memorized but no doubt existed. Insofar as this is true, no one can coincide immediately with his or her own origin. Then the 'fabulation' intervenes into its tear or void at the very core of our existence and functions as a 'supplement of origin' (*supplément d'origine*) – in the Derridean sense, but Bergson himself used this term – for 'inventing' the tie which should attach me to my own origin. For this reason, as many national foundation myths show it, the fabulous story tries to contain its own origin in itself; or else it tries to justify itself by the impossibility or absence of origin.

However, to fill the void with fabulous history is precisely the 'sickness of history', as Ricœur puts it in *Memory, History, Forgetting* (Ricœur, Blamey and Pellauer 2006). Ricœur refers to Nietzsche's reflection on the opposition of 'life' (*Leben*) to 'history' (*Geschichte, Historie*) in *Unfashionable Observations*. Why does history or historicism harm life? Nietzsche employed a very interesting metaphor: insofar as we admit 'life' to serve 'history', we measure the largeness of a tree only from its visible trunk and branches without embracing the whole wood. In other words, 'history' defined as curios (eine *antiquarische* Art der Historie), being far from replacing the individual personality in the vast field, comes to shrink it as well as its vision. 'Contraction' as it were. To cure us of this sickness, Nietzsche recommends that we live against history, unhistorically and transhistorically. And this, in order to save something new, something which *becomes* against both the inevitably eliminatory and integrating process of flatting and homogenizing by history or historiography as curios. It seems crucial to me that Nietzsche names this unhistorical position 'monumental history'. As you know well, the word 'monument' is derived from '*monere*' (*rappeler*, remind). Thus, the breaking or tearing of the tie the 'memory' has with 'history as curios', in other words, an undoing of 'memory as history' and 'historically ordered memory' becomes possible by 'history as memory'.

Severing a tie of memory with history is nothing but 'forgetting', which, in turn, functions as a discovery of the unmemorable, immense field beyond history. It is in this way that flattened and homogenized things recover their status of unique events. For example, Hesiod said in his *Theogony* that Mnemosyne, that is Memory, consists in forgetting of unhappiness, but the forgotten isn't only what one has been conscious of before being forgotten. It also refers to what can be re-evoked by anamnesis, but also what always remains forgotten. So, in this sense, the forgotten isn't the forgotten. Nietzsche, and probably Ricœur himself found there the task of hermeneutics, which is tightly connected with that of ethics (sin, crime, forgiving and so on).

Today I cannot enter these problematics of hermeneutics and of ethics, but one important point to emphasize is the fact that not only Ricœur but also many diverse philosophers or thinkers such as Heidegger, Rosenzweig, Benjamin and Levinas were profoundly inspired by this conception of Nietzsche. I said earlier that Benjamin criticized the absence of historicity in Bergson; nevertheless, Benjamin himself tried the saving (*Erlösung*) of the past as constellation through his critical tearing of '*Jetztzeit*', which makes me think about the '*Lebendige Gegenwart*' in Husserl as well as the '*Augenblick*' in Heidegger. Furthermore, I can't help mentioning the case of Simone Weil. According to Weil, the fabulated attachment of individual to the closed society is precisely its '*déracinement*'; each individual has plural roots or radixes which expand beyond my 'memory as history'. However, if it were the case, I'm obliged to wonder if, in *Time and Free Will* as well as in *Matter and Memory*, Bergson had the same intention to 'reduce', to 'put into bracket' the so-called history. So, what are the consequences of this fifth hypothesis? That is the last question which I will now try to answer.

4

The 'pure memory' testifies that we have an immemorial part in ourselves. It's impossible to fix the beginning of this process, but from the beginning until the formation of 'I' via mirror-stage, we cannot have any memory. Then what happens in this period? Even to this day, nobody knows. I do not think any theory of cognitive development has so far succeeded in explaining the language acquisition process. However, it is clear that such an acquisition is not a simple acquisition. In becoming a speaker of Japanese language, for example, newborn babies lose their capacity to accommodate corporally to other languages. As infants, they live as it were the 'nebula-language' from which plural languages will differentiate. In other words, 'nebula' makes possible mixture or concomitance of incompatibles plans.

To tell the truth, the 'nebula' is a keyword of Charles Renouvier as well as of Bergson, who writes in *Matter and Memory* that 'there is a continuous progress through which the nebula of idea (*nébulosité de l'idée*) condenses into audible images that are distinct from one another, which, in turn, solidify as a coalescence with the materially perceived sounds' (Bergson 1959: 266). It is in this way that what Bergson calls 'language as idol' forms itself. Therefore, according to my sixth hypothesis, what Bergson names 'nebula of idea' should be taken as 'nebula-language' which has not yet been differentiated into languages. It is true that a fluid nebula differentiates and solidifies itself little by little in connection with the contraction of habit; nevertheless, as Bergson himself said in regard to the matter of evolution of life, this process also must have its counter-process of nebulalization; the latter not only has destructive and negative function, but it's precisely the ever forgotten and forbidden, that is the virtual condition for our capacity of speaking.

Thus, what I call nebulalization isn't a simple degeneration at all; if so, the sense of sickness and trouble will change radically. Now acquisition is a kind of loss, loss is a kind of acquisition. As shows it the neologism 'normopathe', sickness has ceased to be the opposite of normal. In Chapter 2 of *Matter and Memory*, Bergson, as you may well know, tried to show how aphasia isn't a result or suppression of memory itself, and he suggests that a patient of linguistic deafness falls into the same state as we do when we hear people who speak an unknown langue. I presume Bergson wrote this passage as a simple comparison, but in fact, it presents the 'reality' for me. Even if his sickness were due to some organs, the patient is obliged to be brought back to the 'nebula-language' because of the tearing of the fabulous tie which associates my body with my langue. In a word, the identificatory relation of the fabulation to the closed society is dissolved or shattered and unheard-of adventure or exploration begins.

Insofar as 'fabulation' is a natural imaginative function of a closed society, what I've said is related to the qualitative distinction that Bergson established between open society and closed society. As for the Christian mystic who comes to open the '*clôture*' of society, Bergson writes in *Two Sources* that 'when the true mystic speaks, there is something which echoes it imperceptibly in most of us, or rather he discovers the marvelous perspective in us' (Bergson 1959: 1157). I can't help but point out the terminological resemblance between this passage and my description of 'pure memory' as 'nebula-language'. However, the relation between 'fabulation' and 'mystic speech' is different from that between 'fabulation' and what I call 'nebula-language'. Although both of them liberate us from the closed society, the former tends to purify whereas the latter is an oversaturated mixture, the former steps outside the '*clôture*' of a society whereas the latter opens this '*clôture*' itself so to speak, the former is a breakthrough by '*élan vital*' whereas the latter is a fluid destruction which often appears as degeneration, the former is modelled on the so-called Christian exceeding of paganisms as well as of Judaism whereas the latter is rather a negation of this hierarchization.

In fact, however purified, the Evangel and the mystics who follow it must be a kind of fabulation or a product of fabulation. Further, if mystic's speech or silence could reach the humanity after Babel at all, it cannot be irrelevant to 'nebula-language'. Or so it seems at least for me. Where Bergson talks about the Christian mystic, Ricœur develops his hermeneutics of 'kerygma'. Here, please allow me to mention the name of Levinas, who considers 'kerygma' as 'the said' (*le Dit*) and goes beyond it to the pre-original 'saying' (*Le Dire*). Monotheistic *polemos*. However, I think not only Ricœur but also Levinas follow, mutatis mutandis, the way opened by *Two Sources* of Bergson, but it should not be forgotten that Bergson himself writes that 'dynamic religion can transmit itself only by image and symbol of fabulation' (Bergson 1959: 1203) and compares the mystic 'outside' to the 'fringe' or 'margin' of intuition. If it were the case, the fabulation wouldn't be a simple opposite to the mystic, and, as I said in my lecture in 2007, the '*élan vital*' would be nothing other than vibration of the *clôture* itself (Goda 2012). In the last chapter titled 'Final Remarks' in *Two Sources*, Bergson comes to point out the 'void between Soul and Body' (Bergson 1959: 1239) and clarifies the supplemental and reciprocal relation between mystical and mechanical. If so, it is possible that Bergson would try to venture into the research on unknown 'law' of dynamic process of what I called above 'becoming-human body.' That could be my final hypothesis.

References

Bergson, H. (1959), *Oeuvres*, Paris: Presses universitaires de France.
Changeux, J.-P. (1985), *Neuronal Man: The Biology of Mind*, New York: Pantheon Books.
Goda, M. (2012), 'Poétique des ruines. Bergson/Benyamin/Levinas', in Shin Abiko, Hisashi Fujita and Naoki Sugiyama (eds), *Dissémination de l'Evolution créatrice de Bergson*, 231–9, Hildesheim: Georg Olms Verlag.
Malabou, C. (2011), *Que faire de notre cerveau?* Montrouge: Bayard, Impr.
Ricœur, P. (1950), *Le volontaire et l'involontaire*, Paris: Aubier.
Ricœur, P. (1960), *Philosophie de la volonté. 2: Finitude et culpabilité*, Paris: Aubier.
Ricœur, P. (2009), *Philosophie de la volonté. 1: Le volontaire et l'involontaire*, Paris: Editions Points.
Ricœur, P., K. Blamey and D. Pellauer (2006), *Memory, History, Forgetting*. Chicago: University of Chicago Press.

5

'A Long-accepted Foreigner, To Whom One Grants Refuge for a While': On Riquier's Interpretation of Bergson's Kantianism

Hisashi Fujita
Kyushu Sangyo University

1 Bergson and Kant (always and again)

1.1 The post-Deleuzean phase

Deleuze, who revolutionized Bergson's studies with his ground-breaking reading, died in 1995. Subsequently, in 1997, Frédéric Worms published his detailed reading of *Matter and Memory*. The period between these two years witnessed the embryonic movement of a post-Deleuzian reading of Bergson. Since then, two decades have passed, and the renaissance of Bergson studies in France, led by Worms, has gradually emerged. The series of studies by Camille Riquier, along with the works of Elie During and others, is proof of this renaissance. In this chapter, Riquier's four articles on Bergson's Kantism (Riquier 2011/2016; Riquier 2011; Riquier 2014; Riquier 2017) will be taken as a comprehensive project to present it as one of his major contributions to Bergson studies.

From Jankélévitch (1959) to Trotignon (1992), from Barthélemy-Madaule's *Bergson adversaire de Kant* (1966) to Philonenko (1994), from Deleuze (1966) to Worms (2001), much has been said about the relationship between Bergson and Kant. Riquier strived to tackle the discussion of this relationship again but with the aim of presenting it as different from animadversion,[1] rejecting the conventional interpretation that has always treated the relationship as oppositional. Although some criticisms were made, Bergson inherited a large part of Kant's basic framework of thought, as observed in the last chapter of the *Creative Evolution*. For him, the problem is not the denial of Kant, but the criticism of Kant. He did not wish for the complete destruction of Kant's philosophy but its transformative acceptance. Even if it was 'a Kantianism in reverse (kantisme retourné)' (Jankélévitch 2015: 39) that ultimately led Bergson in a direction that Kant himself did not adopt, it still proved to be a kind of Kantianism. Let us begin by acknowledging this fact, as it is the starting point of Riquier's argument.

Overall, two possible directions are available for a productive reading of the relationship between the two. While it is interesting to see a certain hidden Bergsonism (*bergsonisme larvé*) in Kant's work (Panero 2008), as attempted by Alain Panero, Riquier attempts to restore a completed Kantianism (*kantisme accompli*) in Bergson. The key term used in the titles of his essays indicates the direction of Riquier's reading: *relève* (recommencement) was adopted by Derrida as a translation of Hegel's *Aufhebung*, and Riquier himself was aware of its usage. 'Bergson's Kantianism will certainly end up betraying the spirit of Kant, but what he means to sublate (*aufheben*) [*relever*] – in the double meaning of to overcome and to complete – through its revitalization in intuition, is no more and no less than Kantian metaphysics, the conditions of the possibility of which Kant had teased out before asserting its impossibility' (R1: 39/65). Then, how does Bergson *relève* a Kantian metaphysics?

1.2 The trap of the history of philosophy: Kant's metaphysics

The history of philosophy can either open our eyes or cover it. Presently, a twofold trap is covering our eyes. On the one hand, there is the trap concerning Kant. While writing the Three Criticisms, Kant dismantled conventional metaphysics. However, he did not want to stay at these 'critical' stages, trying to lay two main metaphysical foundations that had been wiped due to criticism (*Metaphysical Foundations of Natural Science* and *Metaphysics of Morals*). While the division of metaphysical systems must be complete, the complete division of empirical things is impossible, and 'the system itself cannot be expected, but only approximation to it'. For instance, with regards to the concept of right in *Metaphysics of Morals*, 'that right which belongs to the system outlined *a priori* will go into the text (*in den Text*)', while the empirical 'cannot be brought into the system as integral parts of it (*als integrierende Teile*)[2] but can be used only as examples in remarks (*in die Anmerkungen*). . . . otherwise, it would be hard to distinguish what is metaphysics here from what is empirical application of rights' (Kant 2017: 3). The empirical would forever remain outside the text in extra-linguistic notes as exemplification rather than as part of the system, never allowed to set foot on the promised land. Accordingly, the only appropriate title for the first part of *The Metaphysics of Morals* is 'Metaphysical First Principles (*Anfangsgründe*) of the Doctrine of Right'.

The question is whether this pursuit would yield a rich metaphysical future. The framework of Kant's philosophy, driven by fundamental conceptual pairs, such as the sensible and the intellect, the phenomenal and the intelligible, the transcendental and the empirical, the pure and the impure, and the *a priori* and the *a posteriori*, has created a system that is as solid as any, yet as immobile as any future metaphysical inquiry. I would like to take this contribution of criticism seriously. 'To deny that this service of criticism is of any positive utility would be as much as to say that the police are of no positive utility because their chief business is to put a stop to the violence that citizens have to fear from other citizens, so that each can carry on his own affairs in peace and safety' (CPR: Bxxv). Although this notion is true, it is undesirable to have a country wherein thought and action atrophy because the police are extremely vigilant. There is no contradiction between recognizing the positive utility of the police and keeping an appropriate critical distance from them, even engaging in civil disobedience sometimes for the sake of freedom. Greatness in the criticism of old metaphysics is not the same as greatness in constructing of a new metaphysics. 'The criticism of our knowledge of nature that was instituted by Kant consisted in ascertaining what our mind must be and what Nature must be if the claims of our science are justified; but of these claims themselves Kant has not made the criticism' (CE: 379). Still, did Kant's metaphysics not envisage another possible path? Bergson tries to get to the heart of that possibility.

1.3 On intuition

Bergson said that among all the philosophers, Kant was the one who most clearly attributed the success of metaphysics to the reality of intuition. In his article, 'The Perception of Change', published in *Creative Mind*, he wrote: 'one of the most profound and important ideas in the *Critique of Pure Reason* is this: if metaphysics is possible, it is through a vision and not through a dialectic.... He definitively established that, if metaphysics is possible, it can be so only through an effort of intuition' (CM: 164–5). For Bergson's purely sustained intuitive metaphysics, Kant not only dealt a fatal blow to traditional metaphysics but was even the first to *relever* it by using criticism in an astonishing way to overcome any criticism.

However, Bergson immediately adds, 'Only, having proved that intuition alone would be capable of giving us a metaphysics, he added: this intuition is impossible' (CM: 165). What does this statement mean? For the solid metaphysical system described above to be established, intuition must be integrated into the system under close supervision. 'By regarding intelligence as pre-eminently a faculty of establishing relations, Kant attributed an extra-intellectual origin to the terms between which the relations are established.... Thereby he prepared the way for a new philosophy, which might have established itself in the extra-intellectual matter of knowledge by a higher effort of intuition' (CE: 378). Nevertheless, what was deployed within the fortress of criticism was, at any rate, 'sensible intuition' (i.e. either 'empirical intuition' of diverse phenomena, or 'pure intuition' of time and space). The traditional, pre-Kantian 'intellectual intuition' from ancient Greece, which held that it was possible to supersensibly reach the essence of things, was strictly forbidden to enter or leave in the name of the 'critique of reason'.

1.4 The structure of the critique of pure reason

Before comprehensively looking into Bergson's strategy, we should discuss the basics here.

The main part of the *Critique of Pure Reason* is categorized into two parts. Transcendental aesthetic is the study of the rules of sensibility, treating the two pure forms of pure intuition (i.e. space and time) as principles of *a priori* knowledge. Transcendental logic is the study of the rules of the understanding, classified into transcendental analytic and transcendental dialectic. 'Perhaps by oversimplifying Kant's thought,' says Bergson, 'we might say that transcendental analytic demonstrates the congruence of physics, and transcendental dialectic the illegitimacy of metaphysics'

Table 5.1 The divisions of the *Critique of Pure Reason*. © Hisashi Fujita

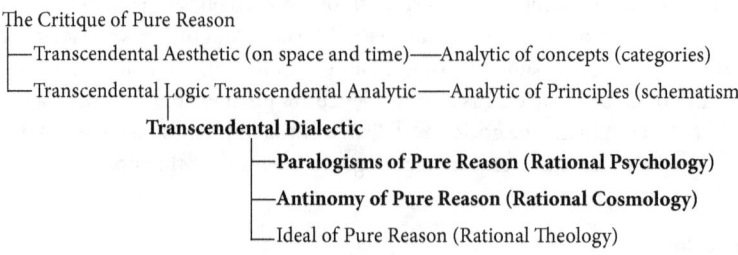

(CIII: 155). Transcendental analytic, which intends to examine and decompose the capacity of the understanding itself, can be further divided into 'analytic of concepts' (so-called Transcendental Deduction) and 'analytic of principles'. The former evaluates and clarifies 'what a concept is', and the latter addresses how that concept is applied to the object of the understanding. How is the application of the concept of the understanding to the object given in time and space by sensibility? If the task of the *Critique of Pure Reason* is to answer the fundamental question, 'How is a synthetic *a priori* judgment possible?' or, in other words, 'How can categories, which are subjective conditions, have objective validity?' On this note, Bergson declares that 'here we are at the very center of Kant's theory' (CIII: 165). Thus, the *a priori* pure concepts of the understanding called 'categories', together with the 'transcendental scheme' that mediates between it and the different empirical intuition, and the 'principle' that defines the legitimate application of the scheme are derived. Therefore, the view of truth transforms as well. 'For ancient philosophers, the truth was a faithful reflection of things into the intuitive intellect, which receives light from things. For Kant, however, truth is the active product of the understanding, and this understanding applies through schemata its *a priori* pure concepts to the data of sensible intuition, data whose origin is the refraction of things-in-themselves in space and time' (CIII: 168–9). Hence, transcendental analytic is termed the 'logic of truth' (CPR: A62) by Kant. What we saw until here is the defensive side of Kantian theory building.

All our knowledge begins with experience, but not all emerge from experience. 'All our cognition starts from the senses, goes from there to the understanding, and ends with reason, beyond which there is nothing higher to be found in us to work on the matter of intuition and bring it under the highest unity of thinking' (CPR: A298). Suppose it is 'transcendental aesthetic' and 'transcendental analytic' that actively elucidate the process of the formation of this cognition and its legitimate possibility and validity. In that case, 'transcendental dialectic' exposes the latent impulse within reason itself to overstep its legitimate limits and delusions. Hence, the offensive side stems from the defensive side. If the transcendental aesthetic discusses sensibility and the transcendental theory of analysis considers the understanding (*Verstand*), the problem of transcendental dialectic is the reason (*Vernunft*). The role of the reason, which has no intuitive form (e.g. sense) and no concept (e.g. the understanding), is to drive the understanding to constantly extend the application of categories farther and farther to introduce as much unity as possible into cognition. 'Just as little as we can avoid it that the sea appears higher in the middle than at the shores,' the illusion in transcendental judgements 'irremediably attaches to human reason, so that even after we have exposed the mirage it will still not cease to lead our reason on with false hopes, continually propelling it into momentary aberrations' (CPR: A297–8). The role of transcendental dialectic is the critique of these dialectical appearances, exposing them in the transcendental judgements and removing the confusion each time to avoid deception. At the core of this transcendental dialectic lies the theory of antinomies.

The antinomies, which belong to the last of the main parts of the existing composition, marks the initiation of the *Critique of Reason*, as confirmed by Kant's own words in his letter to Garve dated 21 September 1798: 'It was not the investigation of the existence of God, immortality, and so on, but rather the antinomy of pure reason ... that is what first aroused me from my dogmatic slumber and drove me to the critique of reason itself' (Kant 1999b: 552). From this point of departure, he searches for a transcendental idea, i.e. a purely rational concept whose conceptual object is not empirically given and is completely outside the capacity of the pure understanding. This transcendental idea, which concerns the unconditional total unity of all conditions in general, is of three kinds. They are the ego (immortality of the soul), the world (freedom), and God. The analysis of the nature of these concepts of pure reason is conducted using categories as a guide. Altogether, three kinds of dialectical reasoning are present, namely 'paralogisms', 'antinomies' and 'the ideal'. They correspond to three traditional forms of syllogism: paralogisms to *categorical* syllogism, antinomies to *hypothetical* syllogism, and ideals to *disjunctive* syllogism (CPR: A335/B392–3, A406/B432–3). Paralogism arrives at the idea of 'ego (soul)' as the 'absolute unity of thinking subjectivity' by means of categorical syllogism (A is B, B is C, therefore A is C). Using hypothetical syllogism (if P, then Q, if Q, then R, therefore P, then R), antinomy gives way to the notion of the world as 'the absolutely unconditioned in a series of given conditions'. Finally, the Ideal derives the idea of God as 'the Being of all beings' via disjunctive syllogism (X or Y, not X, therefore Y).

2 *Aufhebung* of the *Critique of Pure Reason*

2.1 Bergson's path

Kant's successors, Fichte, Hegel and Schopenhauer, failed to challenge Kant's Everest head-on from its summit (i.e. the transcendental dialectic), as stated by Riquier (R1:51/73).[3] They all took the liberty of reclaiming the intellectual intuition that Kant had solemnly forbidden (attributable to the very good reason that it transcends the essentially finite and sensible human condition) and attempted to fly away outside time and space to 'project itself with one bound into the eternal' (CM: 34). However, they fell into the empty notions like Ego, Idea or Will, and in the end only repeated the mistakes of the ancient philosophers even more vividly; while they thought having stepped outside the confines of the intellect, their intuitions were nothing but *intellectual*. Their mistake, according to Bergson, was to follow Kant too closely (contrary to appearances) when drawing a clear separation between the two intuitions. This is why from the beginning of the *Introduction to Metaphysics*, the definition Kant ascribes to 'intellectual intuition' (and subsequently brands as inaccessible and unjustifiable to rational, finite beings) is immediately revisited. To acquire (super-)intellectual intuition, Kant only required sensible intuition.

> There would be an intuition of the psychical, and more generally of the vital, which the intellect would transpose and translate, no doubt, but which would none the less transcend the intellect. ... if we have an intuition of this kind (I mean an ultraintellectual intuition), then sensuous intuition is likely to be in continuity with it through certain intermediaries, as the infra-red is continuous with the ultra-violet. Sensuous intuition itself, therefore, is promoted (*se relever*). ... there are thus two intuitions of different order. ... But this duality of intuition Kant neither would nor could admit. ... Nothing could be more contrary to the letter, and perhaps also to the spirit, of the *Critique of Pure Reason*.
>
> CE: 380–2

To transcend the human experience, there is no other way, and that alone is sufficient: to first place oneself in the midst of one's sensible intuition and then to dive deep into the human experience. Sensible intuition has already touched the absolute, thereby secretly leading to metaphysical and intellectual intuition.

In his endeavour to reach the highest peak, Bergson took the hardest, most circuitous, but most certain path (i.e. from aesthetic through analytic to dialectic). Bergson, who had hesitated to employ the term 'intuition' for a long time, since he knew that it added to the confusion, finally made a resolute decision: to reconfigure the three-part structure of 'criticism' (aesthetic/analytic/dialectic) boldly, by dismantling and reconstructing it in earnest, he was determined to free from the straitjacket his 'intuition', which Kant had allowed his system to accept only under strict confinement.

Among Riquier's four papers, the first one (2011/2016) presents an overall program that painstakingly traces the various strategies for the '*Aufhebung*' of Kantism in the development of Bergson's writings, while the other three papers elucidate respectively each part of what Riquier regards as the central part of the *Critique of Pure Reason*, i.e. transcendental dialectic. The second paper examines the antinomy; the third paper analyses the first three paralogisms; and the fourth paper, particularly the fourth paralogism.[4] According to Riquier, it is only through Kant's critique of rational psychology, cosmology and theology that we can draw a pointillist outline of Bergson's own intuitive metaphysics. Such intuitive metaphysics is 'neither below nor beyond Kantian critique, but can pass through it, can traverse it in its entirety' (R1: 65). There exist many benefits in clarifying Bergson's Kantianism, the first of which is to confer a precise status to Bergson's metaphysics, and the second is that we can clarify the position of Bergson's philosophy in the field of metaphysics (i.e. psychology, cosmology, and theology). After sharply rejecting general metaphysics (ontology) and replacing it with epistemology, Bergson appears to have followed the program of particular metaphysics, starting from psychology, tracing the continuous thread of experience and then exploring cosmology and theology in turn.

2.2 Confrontation with the transcendental aesthetic

It is as if Bergson had made up his mind in advance to spend his life confronting the *Critique of Pure Reason*. His first book, *Time and Freewill*, depicts his confrontation with the transcendental aesthetic, the first gate of the firm fortress of the *Critique of Pure Reason*. 'In this very confusion of true duration with its symbol both the strength and the weakness of Kantianism reside' (TFW: 233). For Bergson, Kant's achievement was making both time and space 'an intuition, not a concept'. While dogmatism distinguished from the understanding to sensibility with the 'degree of clarity' and found only 'attenuated intelligence' or 'confused intelligence' in sensibility, empiricism discerned between the two according to the 'degree of activity' and only perceived 'various representations lacking activity' in the understanding. Unlike both of those, Kant, for the first time, saw the understanding as the faculty of concepts and sensibility as the faculty of intuition, the two being 'deeply distinct' but separately 'original' (CIII: 146).[5] Kant's philosophy depicts time and space as two *a priori* forms of sensibility, which 'would not be any more in us than outside us; the very distinction of outside and inside would be the work of time and space'. Its 'advantage' consists of providing for our empirical thinking 'a solid foundation, and of guaranteeing that phenomena, as phenomena, are adequately knowable' (TFW: 233–4).

Conversely, for Bergson, the doubtful aspect ('legitimacy is far from obvious') of the transcendental aesthetic lies in the 'constant parallelism' (CIII: 151) set up by Kant between time and space. 'Kant's great mistake was to take time as a homogeneous medium' (TFW: 232). As spatial representations of external objects are perceived by us only after they become internal states, and because time is the condition of all representations, 'there is, therefore, a difference in generality between these two forms [of time and space], but not a difference in nature' (CIII: 147). Consequently, duration escapes mathematical consciousness, which retains only simultaneity from time and immobility from movement, and any linkage in the phenomenal world that cannot be translated into simultaneity (i.e. space) becomes scientifically unrecognizable, and genuine psychology remains unrealized.

Nevertheless, both the strength and weakness of Kant's philosophy exist *simultaneously*. If time, like space, had not been regarded as a medium indifferent to what fills it, the 'controlling factor in the whole of this theory', that is, the 'vital distinction' (TFW: 234) between the matter of consciousness and its form and between the homogeneous and the heterogeneous, would probably never have been made. Riquier points out: 'The point is subtle – a spatialized time – but from the perspective of a Kantianism of Bergson, this objection is not what closes the dialogue with Kant, but is its opening' (R1: 50/72). The dialogue continues and deepens, for instance, in *Creative Evolution*: 'What the Transcendental Aesthetic of Kant appears to have established once for all is that extension is not a material attribute of the same kind as others' (CE: 215).

2.3 Confrontation with transcendental analytic

Bergson is fully aware of both a necessity for and an inevitability of the Kantian demand: to move from sensible and empirical intuition to (super)intellectual and metaphysical intuition, we must go through the detour of criticizing the everyday tools of human cognition. That is to say, between aesthetic and dialectic, the analytic must intervene. Bergson's Kantianism has undertaken a major renovation of the various channels of the transcendental analytic to make it consistent with the intuitive experience of pure duration. Riquier adopts two major modifications.

First, Bergson, who believes that the objective validity of mechanics can be secured only by the concept of 'space' inherent in things, no longer needs category tables, neither the category of causality that Schopenhauer adhered to, nor synthetic *a priori* judgements. According to Bergson, Kant mistakenly believed that the transcendental ideality of time and space must be proved to be able to afford them an empirical reality. The formal intuition of space can itself be an object of real experience. Hence, instead of the transcendental deduction, the deepest part of *Critique of Pure Reason*, part almost completely rewritten in the second edition, Bergson in *Creative Evolution* proposes to map out the true genesis of spatiality and intelligence. His concept of 'space' changed its character from 'an *a priori* form of sensibility' (*Time and Freewill* or Lecture on CPR) to '*schème*' (*Matter and Memory*) and finally to '*schéma*' (*Creative Evolution*), which belongs to the sphere of intellect. This transition aimed to shift the concept of space from the aesthetic to the analytic, thereby giving the weight of the constitution of objectivity to the analytic, to render back to sensible intuition a direct access to the things themselves (R1: 52/74).

The second and main modification is that the unity of human cognition's system is no longer based on 'thought' (i.e. on the synthetic unity of apperception), as Kant assumed ('The *I think* must *be able* to accompany all my representations' (CPR: B131)), but on 'action'. This theory of cognition was sketched by Bergson in *Matter and Memory*, enabling us to re-examine the tools of our perception by emphasizing their 'limitation' rather than their 'relativity' (R1: 56/75). If we account for 'a radical and complete empiricism' which Kant seems to repudiate, 'especially the thesis of the evolutionists', there is no need for synthetic *a priori* judgements (CIII: 144–5), for such an empiricism 'would go to the end of its principle, assuming on the one hand that certainty does not differ in its essence from an infinite probability; and on the other hand that experience being infinite, indefinite, extending over generations without number, can produce the equivalent of a transformation of the intelligence' (CIII: 145). The *a priori* will ultimately turn out to be nothing more than habit.[6] Even so, the limitations imposed by Kant in his 'transcendental analytic' are retained. By abandoning the realm of experience for purely speculative uses, the principles of cognition and framework of action create insoluble difficulties: 'the insoluble difficulties into which philosophy falls, the self-contradiction into which the intellect can fall when it speculates upon things as a whole – difficulties and contradictions we naturally come up against if the intellect is especially destined for the study of a part, and if we nevertheless mean to use it in knowing the whole' (CM: 44). Thus, Bergson's famous 'resolution of the false problem' is a continuation of Kant's theory of antinomy.[7]

Due to these two renovations, it becomes possible to reconcile science and metaphysics in the very place where Kant made his sharp demarcation (R1:54/75) because Bergson restores the honour of experience with a dual qualification while intersecting both domains. First, he brings metaphysical problems back into the strictly scientific realm of facts (i.e. analytic). Subsequently, owing to intuition, he gives to the experience an extension that it did not have in Kant's formulation, allowing for the erosion of the domain of metaphysics itself (i.e. dialectic).

This is the point of divergence between Bergson and Kant. For Kant, 'the altered method of our way of thinking' due to the Copernican turn, which is how 'we can cognize of things *a priori* only what we ourselves have put into them' (CPR: Bxviii). The transcendental dialectic aims to exert 'a critique of the understanding and reason in regard to their hyperphysical use, in order to uncover the false illusion of their groundless pretensions and to reduce *their claims to invention and amplification*, putatively to be attained through transcendental principles, to the mere assessment and evaluation of the pure understanding, guarding it against sophistical tricks' (CPR: A64/B88). In short, 'that as mere speculation it [=metaphysics] serves more to prevent errors than to *amplify cognition* does no damage to its value, but rather gives it all the more dignity and authority through its office as censor' (CPR: A851/B879). On the other hand, as Jean-Louis Chrétien, one of masters of Riquier, points out, 'The great singularity of Bergson ... is that dilatation does not form for him an object among others, but the very name of his philosophical method. This is the only example where 'dilatation' designates the central question of philosophizing. For Bergson, dilatation is nothing more than another name for intuition in the new sense that he gives to this term.... The dilatation begins with the will, it is an act and an effort, and this effort that expands our consciousness and our time transforms the capacities of our intelligence' (Chrétien 2007: 26–7). In Bergson's philosophy, the transcendental dialectic, accompanied by the twofold demand mentioned previously (i.e. the positivization of metaphysics and the extension of experience), ceases to demarcate the perimeter of an impossible metaphysics and gradually transforms itself into a concrete metaphysics.

2.4 Traps in the history of philosophy again: Bergson's 'empirical metaphysics'

At the beginning of this chapter, Kant's 'metaphysics' is presented as a trap set specifically for the history of philosophy. The other trap concerns the meaning of 'in reverse' in Bergson's 'Kantianism in reverse'. Deleuze's reading of Bergson had the common characteristics, immersed in the atmosphere of the epoch, with, for example, Lacan's structuralist reading of Freud – the sharp rejection of all psychology (American ego-psychology for Lacan), and the excessive emphasis of the 'ontology of the virtual' (ontologization of the structure of the unconscious for Lacan) – although this tendency led Deleuze to a drastic decrease of confrontations with various empirical facts of natural sciences.[8] Now, was it not Kant in the *Critique of Pure Reason* who said that 'the proud name of an ontology ... must give way to the modest one of a mere analytic of the pure understanding' (CPR: A246/B303)? By keeping this relationship with Kant in mind, Riquier says, a somewhat difficult passage in the Introduction to *Creative Mind* becomes easier to understood: 'It is, therefore, already fulfilling half of the program of the old metaphysics: it could be called metaphysics did it not prefer to keep the name of science' (CM: 30). What is the meaning of this statement, which is often skipped over in favour of the subsequent assertion that 'They [=science and metaphysics] both bear upon reality itself. But each one of them retains only half of it'? What Bergson calls 'the program of old metaphysics' actually refers to the Wolffian division between *metaphysica generalis* and *metaphysica specialis*, which Kant faithfully followed in the *Critique of Pure Reason*. With Bergson, natural science may have rediscovered the old name of general metaphysics. However, it is only after Kant who replaced 'the proud name of an ontology' with 'the modest one of a mere analytic of the pure understanding' that Bergson allows science itself to carry out one of the two halves of the program of old metaphysics or *metaphysica generalis*.

2.5 Confrontation with transcendental dialectic

Bergson, who 'relativizes relativism and criticizes criticalism', to borrow Barthélemy-Madaule's expression (1966: 59), strictly retains from transcendental dialectic the part of what seems to be always a valid criticism to metaphysics: antinomy among others. 'So that if we claim to affirm something of it, at once there rises the contrary affirmation, equally demonstrable, equally plausible. . . . Such [theory of antinomy] is the governing idea of the Kantian criticism. It has inspired Kant with a peremptory refutation of 'empiricist' theories of knowledge. It is, in our opinion, definitive in what it denies' (CE: 216). Then, to move from sensible and empirical intuitions to (super-)intellectual and metaphysical intuitions, we require the condition that our cognition is 'limited' but not 'relative'. Being relative means being completely separated from the absolute, leaving us forever outside of 'things-in-themselves', or realities. If, on the other hand, we are only limited, it means that we are still held in the 'reality' by our sensible cognitions, even if we can only catch a glimpse of the real.[9] Such metaphysics does not transgress experience; on the contrary, it transgresses the fixed and immutable framework of the understanding to respect 'integral experience' (R1: 56/76).

There are countless scholars who wish to read an ontological agenda into Bergson's philosophy. Deleuze among those claims that Bergson, by elaborating an 'ontology of duration' and creating his own problems, could free himself from the old problems believed to be eternal – the problems of the soul, the world, and God. However, according to Riquier, it is quite the opposite that must be admitted. From an epistemological point of view, not only is there no ontology in Bergson,[10] nor can there be, but also if there exist a metaphysics in Bergson, it can only be the *metaphysica specialis* of the soul, the world, and God that Deleuze wanted to exorcize. It is well known that the *Critique of Pure Reason* made the thinking subject (i.e. the soul) the object of 'psychology', the sum of all phenomena (i.e. the world) the object of 'cosmology', and the Being containing the first condition which makes all conceivable beings possible (i.e. God) the object of 'theology'. Transcendental dialectic depicts the philosophical route wherein pure reason gives ideas to transcendental psychology (*psychologia rationalis*), transcendental cosmology (*cosmologia rationalis*), and the transcendental cognition of God (*theologia transcendentalis*) (A334–5/B391–2).

Table 5.2 Correspondence between Bergson's works and *Critique of Pure Reason* (Riquier). © Hisashi Fujita

TFW	(I) Rational psychology: Paralogisms of pure reason	(1) Paralogism of substance
		(2) Paralogism of simplicity
		(3) Paralogism of personality
		These three concepts give us 'spirituality' (CPR: B403)
MM		(4) Paralogism of the ideality (of outer relation): Its status differs from the first three paralogisms in that it relates the soul 'to objects in space' and bears upon 'the *commerce* with the body'
	(II) Rational cosmology: Antinomies of pure reason	(1) First antinomy of pure reason
		(2) Second antinomy of pure reason
CE		(3) Third antinomy of pure reason
		(4) Fourth antinomy of pure reason
TSMR	Rational theology: The 'Ideal' of pure reason	

In Riquier's image, Kant was not akin to Columbus who actually entered the New World, but Moses who, before entering it, stayed on the borderline, pointed to the *terra incognita*, and then died. In Heideggerian terms, Kant might have flinched at his bold discovery and buried it himself. Commentators will attempt to find something in the third critique, the *Critique of Judgement*, that mediates the first two, but this will take place sometime later. At the time, as Peguy recalls, 'Kantists were naturally divided into two groups: those who thought according to the *Critique of Pure Reason*, and those who lived according to the *Critique of Practical Reason*'. Aside from the confrontation with Kant's morality in the first chapter of *Two Sources*, Bergson, like Heidegger, belongs to the camp of those who have thought through the first critique, the *Critique of Pure Reason*, and his thinking respects the outline presented by it. Bergson's writings follow in Kant's footsteps, reforming it from within, and attempting to carve a way that can lead us anew to metaphysics. The table below is appended to the end of Riquier's first paper (R1: 59/78–9). This table illustrates his strategy well, according to which Bergson's itinerary reflects the constitution of the *Critique of Pure Reason*.

3 Concluding remarks

Although Riquier's formulation of the problematic is bold and circumspect and his argument is sufficiently persuasive, there remain some problems. I would like to outline four points.

The first is the importance of Descartes in *Time and Freewill*. 'In *Time and Freewill*, Bergson had assumed Descartes' legacy' (R4: 44), while ultimately, 'Bergson would have been the same without Kant' (R1: 37/64). This is as if the atmosphere of the epoque obliged Bergson to insert some confrontations with Kant in his works and the Descartes case was fundamentally different from that of Kant. Bergson argues, it is true, that even great ideas can be indeed, at first sight, correlative to the period in which the philosopher lived, although this assumption often exists only on the surface: 'If Spinoza had lived before Descartes he would doubtless have written something other than what he wrote, but that given Spinoza living and writing, we were certain to have Spinozism in any case' (CM 93). Yet is it not the consequence of this argument that we should treat the influence of Descartes and that of Kant in the same way? Riquier, who knows well the formation of *Matter and Memory*, must admit that the central chapter of *Time and Freewill* is the second chapter, which discusses the concept of 'duration', and that Kant's transcendental aesthetic plays a decisive role in it. It may be the influence of the overly powerful reading grid of Jean-Luc Marion, another master of Riquier.

Second is the importance of Kant in *Matter and Memory*. Riquier is of the view that the first and fourth chapters of this book belong to the Kantian problematic, while the second and third belong to the Cartesian problematic: 'The middle two chapters of *Matter and Memory* are devoted exclusively to defining the relation between body and mind in terms of memory, and in this way, solving the problem of union that Descartes traditionally left to him' (R4: 50). As for the fourth chapter, I would like to express my full admiration for Riquier's remark: 'it seems to us that Kant is again the one with whom Bergson continues to argue consistently'. However, I personally think that the two central chapters also attempt a thoroughgoing confrontation with Kant.[11]

The third is the relation between Bergson's philosophy and transcendence. Riquier appears to state that Bergson's extension and expansion of Kant's transcendental philosophy could maintain without altering its original transcendental character; I wonder, however, if this is true. While I have attempted elsewhere to clarify the difference between these two with concepts like 'immanent aesthetic' or 'hyper-schematism', what does Riquier think about this point? For instance, for Kant, 'rational psychology' comprises one of the four main divisions of the whole system of metaphysics, whereas 'empirical psychology' is not granted a permanent residence in the domain of metaphysics.

> Empirical psychology must thus be entirely banned from metaphysics, and is already excluded by the idea of it. Nevertheless, in accord with the customary scholastic usage one must still concede it a little place (although only as an episode) in metaphysics, and indeed from economic motives, since it is not yet rich enough to comprise a subject on its own and yet it is too important for one to expel it entirely or attach it somewhere else where it may well have even less affinity than in metaphysics. It is thus merely a long-accepted foreigner to whom one grants refuge for a while until it can establish its own domicile in a complete anthropology (the pendant to the empirical doctrine of nature).
>
> CPR, A848–9/B876–7

Yet, for Bergson's philosophy, 'empirical psychology' is not a complete 'stranger', and carefully tailored anthropology is not completely foreign. This point should constitute a decisively important issue in considering the relationship between Bergson and Kant.

Lastly, there is the so-called problem of rewriting the *Critique of Pure Reason*. Both the 'transcendental analytic', which Riquier examines in his first paper, and 'the fourth paralogism', which he examines in his fourth paper, have been largely rewritten in the second edition of the *Critique of Pure Reason*, published seven years after the publication of the first edition (1781). In the preface of the second edition, he declared solemnly: 'I have found nothing to alter either in the propositions themselves or in their grounds of proof, or in the form and completeness of the book's plan' (Bxxxviii). However, as we know, he made various important changes. The most important point of contention is the place to give for the imagination, and consequently, the position of transcendental schematism.[12]

As I mentioned at the beginning of this chapter, the post-Deleuzian phase of Bergson studies, as well as Riquier's research, is accelerating and is in a transitory state more than ever. We, too, should now avoid being 'spectators-critics' and start acting like 'actors-researchers'.

Notes

1 Employed by Jankélévitch in his preface to Barthélemy-Madaule's *Bergson as Kant's Adversary*, the word 'animadversion' is rather uncommon: it is a combination of *anima* (mind) and *advertere* (to turn towards), meaning (1) reproach, antipathy, hostility, (2) reprimand [in writing] (by public authority): 'Kant is his eternal enemy (*adversaire de toujours*). From his *Time and Freewill* to his *Two Sources of Morality and Religion*, Bergson has always remained true to himself in this animadversion. It is necessary to inquire into the reasons for this occurrence, asking whether Bergson had sometimes misjudged his own consanguinity in Kant's philosophy by attempting to see in it only an idealistic renunciation of all direct grasping of the real, which is exactly what Madame Madaule did' (Barthélemy-Madaule 1966: vii).
2 It is not impossible to recall and contrast Bergson's 'integral experience'. We shall return to this point at the end of this chapter.
3 As for the first article (Riquier 2011/2016), we note two paginations from French edition, then from English translation.
4 In this chapter, we concentrated on the first and forth paper.
5 As we all know, the two are not equal according to Kant. 'Sensibility, subordinated [*untergelegt*] to understanding, as the object to which the latter applies its functions, is the source of real cognitions. But the same sensibility, insofar as it influences the action of the understanding and determines it to judgments, is the ground of error' (CPR: A294/B351). On the contrary, Bergson argues: 'Sensitivity is not, according to Kant, pure receptivity, pure passivity. It must have its form, its pure form, that is to say a certain original tendency, a certain activity of a certain kind. ... and this sensibility has already organized the objects in its way, bent to its form. It is in this form that the understanding must accept them' (CIII: 146). Egawa's suggestion that sensibility performs a certain passive synthesis (Egawa 2000: 331–3) is consistent with our interpretation of finding an 'immanent aesthetic' in *Time and Freewill*.
6 The idea was still in a vague state in 1889: 'What leads to misunderstanding on this point seems to be the habit we have fallen into counting in time rather than in space' (TFW: 78). When did it begin to take a definite form? We estimate shortly before *Matter and Memory* (1896) as Riquier suggested, around the epoque of 'Lectures on the *Critique of Pure Reason*' (1893–4): 'The empiricist will answer that history must be

taken into account.... The alleged forms of time and space could well be only habits' (CIII: 152). It is clear that this basic idea leads to the overcoming of the transcendental deduction in *Creative Evolution*. 'The habit, according to the opinion of the empiricists of the last century, the heredity according to the opinion of the evolutionists of this century can make that this knowledge acquired by the experience, by consolidating itself, by organizing itself, arrive at being an integral part of the intelligence, take on this character of necessity and of strict universality that Kant holds for primordial, for irreducible. Thus, between the knowledge that Kant calls inductive and which would present only a relative universality and the knowledge that Kant calls a priori and whose universality would be rigorous, there would be only a difference of degree.... Evolutionism makes us witness the genesis of the so-called a priori concepts. An explanation of this kind would not count in the eyes of Kant' (CIII: 138).

7   Deleuze had aptly pointed this out before Riquier. 'Bergson borrows an idea from Kant although he completely transforms it: It was Kant who showed that reason deep within itself engenders not mistakes but *inevitable* illusions, only the effect of which could be warded off. Although Bergson determines the nature of false problems in a completely different way and although the Kantian critique itself seems to him to be a collection of badly stated problems, he treats the illusion in a way similar to Kant' (Deleuze 1991: 20–1).

8   'Strictly speaking, the psychological is the present. Only the present is "psychological"; but the past is pure ontology; pure recollection has only ontological significance' or 'He describes in this way the *leap into ontology*. It is a case of leaving psychology altogether. It is a case of an immemorial or ontological Memory', and to finish, 'in *Matter and Memory*, psychology is now only an opening onto ontology, a springboard for an "installation" in Being' (Deleuze 1991: 56, 57, 76). In terms of Bergson's ontology, see (de Lattre 1990). The fundamental differences between Bergson and Deleuze are explored in my essay (Fujita 2013).

9   The discussion is more detailed in 'Le parallélisme psycho-physique et la métaphysique positive' than in *Creative Evolution* (R1: 56/76–7, M: 494, EC: 357).

10 While I completely agree with Riquier's claim that Deleuzian ontologizing is hermeneutic violence, his stark rejection of ontology also goes too far, because it is precisely the inseparability of epistemology and ontology that characterizes Bergson's philosophy. In the following essay, the quasi-concept of 'hauntology' is used to propose an interpretation different from both Deleuze and Riquier.

11 Without going into details, let us note at least that *Matter and Memory* seems for me to be a work thoroughly thought out in a confrontation with Kant's schematism.

12 In his lecture on the *Critique of Pure Reason*, Bergson states, 'We could thus say, to interpret Kant's thought, that pure imagination is the understanding taking the form of memory' (CIII: 162). In his postface to this lecture, Egawa proposes a truly thrilling interpretation: 'The fact that in Kant, the pure concept becomes applicable to empirical intuition under the schematism of imagination is exactly the same as the fact that in Bergson, the virtual past becomes itself actualized in the actual present' (Egawa 2000: 336). But neither has provided any further elaboration. I have offered my own interpretation in the following article (Fujita 2023).

References

Barthélemy-Madaule, M. (1966), *Bergson Adversaire de Kant, Étude Critique de la Conception Bergsonienne du Kantisme, Suive d'Une Bibliographie Kantienne*.

Bergson, H. (1911), *Matter and Memory*, trans. N.M. Paul and W.S. Palmer, London: George Allen and Unwin, https://archive.org/details/in.ernet.dli.2015.506437/page/n7

Bergson, H. (1911), *Creative Evolution*, trans. A. Mitchell, London: Macmillan, https://archive.org/details/in.ernet.dli.2015.94170/page/n5

Bergson, H. (1946), *The Creative Mind*, trans. M.L. Andison, New York: Philosophical Library, https://archive.org/details/in.ernet.dli.2015.223138/page/n5

Bergson, H. (1972), *Mélanges*, Paris: Presses universitaires de France.

Bergson, H. (1995), *Cours III: Leçons d'histoire de la philosophie moderne. Théories de l'âme*, Paris: Presses universitaires de France.

Chrétien, J.-L. (2007), *La joie spacieuse: essai sur la dilatation*, Paris: Les Éditions de Minuit.

Deleuze, G. ([1966] 1991), *Bergsonism*, trans. H. Tomlinson and B. Habberjam. New York: Zone Books.

de Lattre, A. (1990), *Bergson, une ontologie de la perplexité*, Paris: Presses universitaires de France.

Egawa, T. (2000), 'Bergson's Lecture on Kant: On the Boundaries Between Passive Genesis and Dynamic Genesis', in *Bergson's Lectures*, Vol. III, trans. M. Goda and T. Egawa, 330–43, Tokyo: Hosei University Press.

Fujita, H. (2023), 'Sublime and Panoramic Vision. On Transcendental Schematism in Bergson, Kant and Heidegger', *Bergsoniana*, no. 3.

Fujita, H. (2013), 'The Crossroads of the Philosophy of Life: Metaphysics, Science and Politics in Bergson and Deleuze', in O. Kanamori (ed.), *Epistemology: A History of French Scientific Thought in the 20th Century*, 323–407, Tokyo: Keio University Press.

Jankélévitch, V. ([1959] 2015), *Henri Bergson*, trans. N.F. Schott, Paris: Presses universitaires de France.

Kant, I. (1996), *The Metaphysics of Morals*, Cambridge: Cambridge University Press.

Kant, I. (1999a), *Critique of Pure Reason*, Cambridge: Cambridge University Press.

Kant, I. (1999b), *Correspondence*, Cambridge: Cambridge University Press.

Panero, A. (2008), 'Kant, précurseur manqué de Bergson?' *Revue philosophique de la France et de l'etranger* 133 (2): 133–45.

Philonenko, A. (1994), *Bergson ou De la philosophie comme science rigoureuse*, Paris: Cerf.

Riquier, C. (2011), 'Le jaillissement du monde. La relève bergsonienne des antinomies de la raison pure', in M. Fœssel et al. (eds), *Faire monde. Essais phénoménologiques*, Paris: Mimesis.

Riquier, C. (2013), 'La relève intuitive de la métaphysique: le kantisme de Bergson', in F. Worms and C. Riquier (eds.), *Lire Bergson*, 35–59, Paris: Presses universitaires de France.

Riquier, C. (2014), 'L'entrelacs des traditions. Bergson ou le *cogito* surmonté: les paralogismes de Kant', in A. Panero et al. (eds), *Bergson professeur*, 31–46, Louvain-la-Neuve: Peeters.

Riquier, C. (2016), 'The Intuitive Recommencement of Metaphysics: Bergson's Kantianism', *Journal of French and Francophone Philosophy*, Vol. XXIV (2): 62–83.

Riquier, C. (2017), '*Matière et mémoire* et la relève intuitive de la métaphysique', 37–58, in *Diagnosis of Bergson's* Matter and Memory, Tokyo: Shoshi-Shinsui.

Trotignon, P. (1992), 'Bergson, lecteur et critique de Kant', *Cahiers Eric Weil* vol. III: 'Interprétations de Kant', textes recueillis par Jean Quillien et Gilbert Kirscher, 93–105, Lille: Presses Universitaires de Lille.

Worms, F. (2001), 'L'intelligence gagnée par l'intuition?' *Les études philosophiques*, no. 59: 453–64.

Part Two

Perception and Embodiment

6

Bergson, Gibson and the Image of the External World

Stephen E. Robbins

The hard problem is the image

I must begin with something in John Searle's 2015 book, titled, *Seeing Things as They Are: A Theory of Perception*. In it he notes, '... since the seventeenth century, I do not know of any Great Philosophers who even accepted Naïve or Direct Realism' (Searle 2015: 22). To him, there are only seven 'Great Philosophers', and only these are worth considering on the subject – Kant, Leibniz, Hegel, Descartes, Berkeley, Spinoza and Hume. However, since we are speaking of seeing, a 'greatness' designation is a function of one's ability to *see* greatness. Searle's list is a sad commentary on the state of philosophy. Missing, firstly, is Henri Bergson. Missing, secondly, his complement, the great philosopher-scientist of perception, J.J. Gibson.

The problem is that Bergson's theory has never been understood. Philosophy is totally unaware that Bergson had a solution to Chalmers' famous 'hard problem', and further, that it is a *unique* solution, uncategorized, found nowhere in the supposedly exhaustive list of positions on perception Searle provides, namely, Representationalism, Phenomenalism and Idealism. Searle thinks he is defending a fourth position, namely, Direct Realism. His entire 'solution' relies on the rather discredited notion of *emergence*, that *somehow*, while watching the coffee cup on the table in front of us, with our spoon stirring the coffee, the image of this scene emerges from the biological processes in the brain. However, this image, per Searle, is simply 'in the head'. He does triple flips in his efforts to show that he is not, in reality, an *indirect* realist – that is, as indirect realists hold, we are only seeing this 'image in the head', not the actual objects – the cup, the spoon, the table – where they are, in front of us, 'out there', within the external world. Searle, in truth, has absolutely no theory of seeing. He desperately needs the greatest of the direct realists, Bergson, and also – completely unmentioned in his book – J.J. Gibson.

The 'hard problem,' as formulated by David Chalmers, was roughly this: how, given any neural or computer architecture, does this architecture explain the *qualia* of the external world? Here *qualia* refers to the 'whiteness' of the coffee cup, the 'silveriness' of the spoon, the 'clinking' of the spoon against the side of the cup, but this formulation has been extremely misleading. It has focused philosophy entirely on addressing the origin of qualia. This is so much so that when a theorist (like me) says he has (or at least knows of) a solution to the origin of the image of the external world, philosophers have no idea what he is talking about or why it is important.

This is the problem: what has been missed is that *form* is equally *qualia*, particularly obviously when considered in its dynamic aspect – rotating cubes, buzzing flies, falling, twisting leaves. Listen to Valerie Hardcastle's list of *qualia*: '... the conductor waving her hands, the musicians concentrating, patrons shifting in their seats, and the curtains gently and ever-so-slightly waving ...' (Hardcastle 1995: 1). In other words, she is entirely pointing to both time and dynamic forms, but forms fully populate our image of the external world – the cup, the stirring spoon, the table, the swirling coffee surface. In other words, *everything* in the image is *qualia*. The totality of the image is *qualia*. It is the origin of the image of the external world that is the more general problem. This is what must be explained.

Bergson's model of the origin of the image

The creation of an image is an *optical* problem. It is actually a problem of physics. Though in this case, it is equally a problem of physics in regard to its concept of time. Bergson had a solution that covered both.

The key is in a passage is in his 1896 work, *Matter and Memory*. Bergson had just noted that there can be nothing like a 'photograph' of the external world developed in the brain. We will find nothing remotely looking like the coffee cup and spoon inside the skull. We have seen this ever more clearly in the subsequent findings of neuroscience. But Bergson went on:

> But is it not obvious that the photograph, if photograph there be, *is already taken, already developed in the very heart of things and at all points in space*. No metaphysics, no physics can escape this conclusion. Build up the universe with atoms: Each of them is subject to the action, variable in quantity and quality according to the distance, exerted on it by all material atoms. Bring in Faraday's centers of force: The lines of force emitted in every direction from every center bring to bear upon each the influence of the whole material world. Call up the Leibnizian monads: Each is the mirror of the universe.
>
> MM: 31–2, emphasis added

This was certainly, to his contemporaries, one of Bergson's obscure passages. With the benefit of intervening events, it becomes clear. Fifty years before Gabor's 1947 discovery, Bergson had already envisioned the essence of holography. Let me review this phenomenon for the sake of clarity in what is to follow.

Holographic reconstruction

A hologram is a photographic plate on which the interference pattern of two light waves is recorded (see Figure 6.1). One wave (the 'reference' wave) is a wave of laser

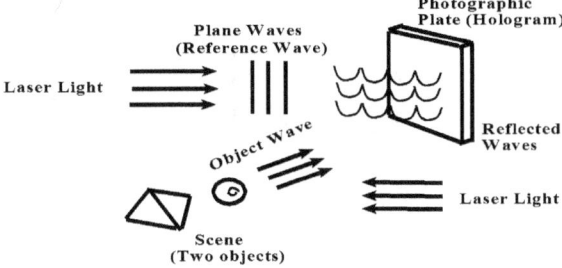

Figure 6.1 Construction of a Hologram. © Stephen E. Robbins.

light that is directly beamed on the plate. The other wave (the 'object' wave) is bounced off an object or objects, say a pyramid and ball, and also covers the plate. The two waves meet at the plate forming an interference pattern. This pattern looks nothing like the original scene, in this case, the pyramid and ball (cf. Kock 1969; Caufield and Lu 1970).

The 'reconstructive' wave is a wave of the same frequency as the original reference wave, we'll say, 'frequency 1' (see Figure 6.2). When the reconstructive wave is beamed through the plate, a viewer now sees the pyramid and ball located in space as 3-D objects. The reconstructive wave is now 'specific to' or specifies the pyramid and ball, i.e. the pyramid/ball as the source of the original reflected wave front.

Using a different frequency of reference wave, say 'frequency 2', we can beam the wave off yet another object, say a cup. Now we 'modulate' the reconstructive wave to frequency 2 – we see the cup (Figure 6.3, right). Modulate back to frequency 1 – we see the pyramid and ball. Thus, the wave fronts from many objects can be recorded on the plate, each via a different frequency of reference wave, and by *modulating* the reconstructive wave appropriately, from frequency to frequency, we can reconstruct each object's original wave front (or image).

Lastly, we should note this: we can consider each point of the illuminated object as giving rise to a spherical wave which spreads over the entire hologram plate. As the

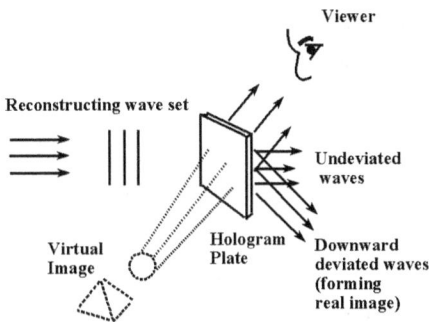

Figure 6.2 Reconstructing the image (i.e. the source of the original wave front). © Stephen E. Robbins.

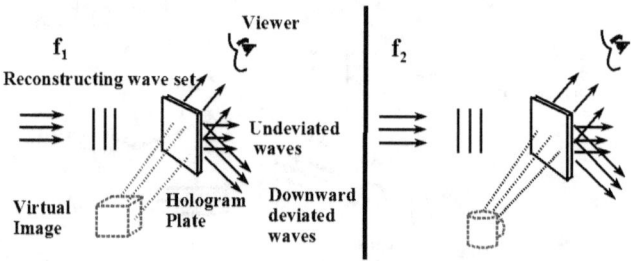

Figure 6.3 Modulating the reconstructive wave – From f_1 to f_2. © Stephen E. Robbins.

reflected light bounced off the pyramid, for example, each point of the pyramid gave rise to a little spherical wave that expanded as it travelled towards the hologram plate, ultimately covering the entire plate. Thus, we can consider the information for each point on the pyramid to be spread over the entire hologram. This has a converse, namely that *the information for the entire object is found at any point in the hologram*. At each point of the hologram is the information for the whole. A tiny corner of the hologram plate can be used to reconstruct the whole pyramid and ball, or the cup, or any of the objects recorded.

Extending holographic principles

This principle can be extended to the entire universe when viewed as a sea of waves, i.e. the universal field can be considered a vast interference pattern – a hologram. Bohm did so in 1980, in his *Wholeness and the Implicate Order*, and this is precisely where Bergson had already gone. Bergson had envisioned the universe as a holographic field (Robbins 2000, 2006a, 2014). This was his '… *photograph already developed in the very heart of things and at all points in space*'; but while Karl Pribram, in 1971, had argued that the *brain* is a form of hologram, Bergson had the correct model. He saw the brain as effectively creating the *modulated reconstructive wave* passing through this external holographic field. This brain-created wave is 'specific to' a source, i.e. a subset of the vast information within the field, and by this process now an 'image' of a portion of the field – our kitchen table, the coffee cup and the stirring spoon (Figure 6.4).

Hard after the 'photograph' passage, Bergson noted:

> Only if when we consider any other given place in the universe we can regard the action of all matter as passing through it without resistance and without loss, and the photograph of the whole as translucent: Here there is wanting behind the plate the black screen on which the image could be shown. Our 'zones of indetermination' [organisms] play in some sort the part of that screen. They add nothing to what is there; they effect merely this: *That the real action passes through, the virtual action remains.*
>
> MM: 32

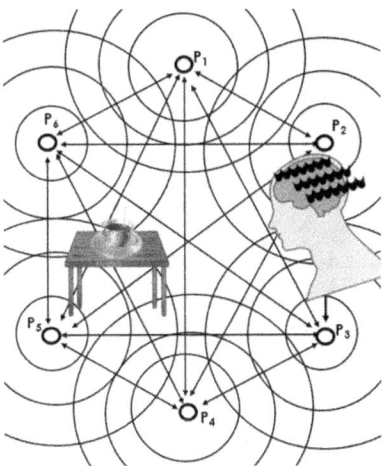

Figure 6.4 The brain acts as a reconstructive wave, passing through the holographic field, specifying a source within the field. © Stephen E. Robbins.

In Bergson's terms, the universal field is a vast field of 'real actions' (one can read 'waves', for concreteness) rippling everywhere – a vast interference pattern. Any given 'object' acts upon all other objects in the field and is in turn acted upon by all other objects. It is in fact obliged:

> to transmit the whole of what it receives, to oppose every action with an equal and contrary reaction, to be, in short, merely the road by which pass, in every direction the modifications, or what can be termed *real actions* propagated throughout the immensity of the entire universe.
>
> MM: 28

The subset of these actions (or information) that the brain-supported reconstructive wave picks out is a portion related to the body's action. This action-relatability is the information-selection principle from the 'hologram'. Thus, *perception*, as Bergson argued, *is virtual action*. We are seeing how we can act. The brain is not 'generating' an image; it is not generating 'experience'. The image, as a specification of a dynamically changing subset of the field, is *within the external field*, right 'where it says it is', not 'in the brain'. This is the ultimate in 'externalism' (though never noted); the ultimate in direct (though far from naïve) realism.

Gibson – the information for modulation

For Gibson too, there is no 'image' being generated 'in the brain', but Gibson did not actually explain how the image comes about. We must place him in the holographic context of Bergson for his theory to truly make sense. For Gibson, there is information

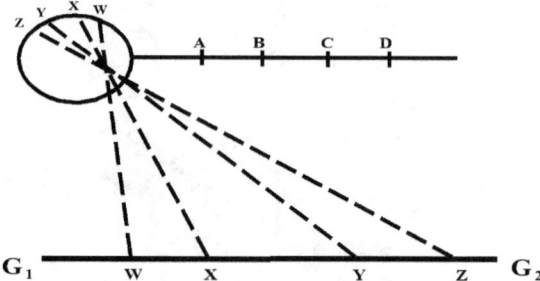

Figure 6.5 The 'ground'. © Stephen E. Robbins.

in the environment 'specific to' – in his terms – the structure of the environment. The brain – yes, like a *wave* – is *resonating* with this information. Therefore, the brain itself is, yes, 'specific to' the environment. Sound familiar?

As an example of this information, Gibson held that the ground surface naturally 'specifies' distance. Berkeley had argued that the points A, B, C, D on the line of Figure 6.5 project to the same point on the retina, therefore, he stated, there is no information for depth. However, Gibson argued that the points on the ground, W, X, Y, Z, via a projective transformation, preserve the same relative distances on the retina. There is indeed information 'specific to' distance.

The ground surface is even more highly structured. There is a texture gradient of texture units (circles in Figure 6.6). These could be little rocks, grains of sand on a beach, a field of grass. The horizontal distance (S) between each unit is inversely proportional to the distance from the observer, that is, the law, $S \propto 1/d$ (d = the distance from the observer). The vertical separation is by the law, $S \propto 1/d^2$. If I move the cup back and forth, its height appears constant. Why? As I move the cup at the back towards the front, there is a constant ratio – the height of the cup (S) increases proportionally as the number of rows (N) of occluded texture units decreases, from 4 rows (rear) to 2 rows (front position) or $S \propto 1/N$. Note the relation to action. This ratio is information

Figure 6.6 Texture gradient. © Stephen E. Robbins.

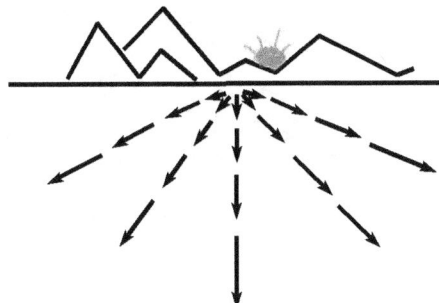

Figure 6.7 Flow field – a gradient of velocity vectors. © Stephen E. Robbins.

used for modulating the hand to grasp the cup in motion. It is the information underlying *virtual action*.

When this surface is put in motion, it becomes a flow field of velocity vectors (Figure 6.7). The value of each vector is $v \propto 1/d^2$. The fastest moving vectors are closest to the eye. The slower moving vectors are further away. There is a ratio defined over this flow termed *tau*. It specifies *severity of impending impact* (Kim, Turvey and Carello 1993). It too is used for guiding action, for example a bird or a pilot uses the *tau* value to modulate or control his landing for a gentle touch down (Figure 6.8).

These flow fields specify form. Figure 6.9 is a 'Gibsonian cube'. As a side turns towards you, there is an expanding flow field. As a side turns away, there is a contracting flow field. The top is a radial flow field. The 'edges' and 'vertices' of the cube are now simply sharp breaks – discontinuities – in these flows.

This makes the specification of form very 'dicey'. There is an inherent *uncertainty* involved in the brain's processing of these velocity flows that I will not go into here (Robbins 2004), but consider the velocity vectors defining the perimeter of a rotating ellipse (Figure 6.10). If the ellipse rotates too quickly, the speed of the motion breaks a constraint ('motion is slow and smooth') used by the brain in its processing. The ellipse

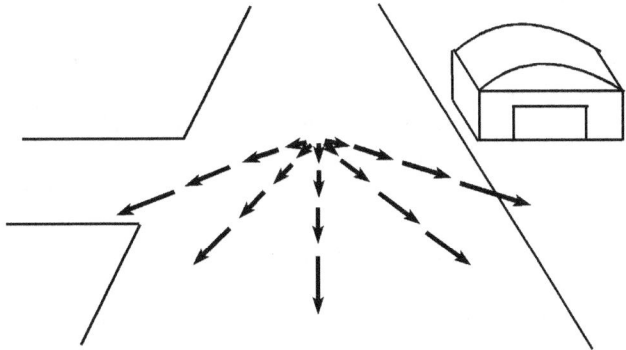

Figure 6.8 Flow field – *tau* value for landing. © Stephen E. Robbins.

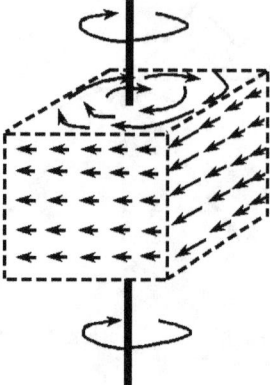

Figure 6.9 The Gibsonian cube – a partitioned set of flow fields. © Stephen E. Robbins.

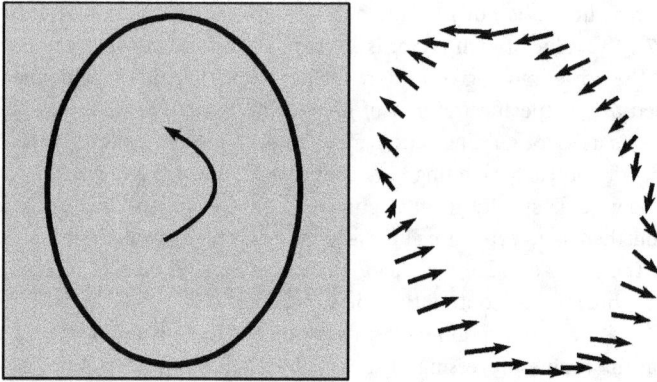

Figure 6.10 Rotating ellipse – velocity vectors. © Stephen E. Robbins.

loses its rigidity; it becomes a wobbly, distorting figure. This is 'Mussati's illusion' (Weiss, Simoncelli and Adelson 2002).

Take a cube made of wire edges (see Figure 6.11). Let it rotate. The cube has a symmetry period of 4, i.e. it is carried or mapped onto itself 4 times in a full 360 degree rotation. If we strobe the cube (with a strobe light) at an integral multiple of this period – 4, 8, 12 times per full rotation, it is seen as a cube rotating. If strobed out of phase – 5, 9, 13 times per rotation – it is seen as a distorted, plastically changing, non-rigid, wobbling object – definitely not a cube.

The specification of the external field then is always an *optimal specification*, that is, a probabilistic specification, based on the inherent uncertainties in determining the velocities of these flows, and therefore on the best information available (constraint broken in the deforming cube case: possibly that 'spatial symmetry implies temporal symmetry'). If I take a reconstructive wave containing frequency 1 and frequency 2, and

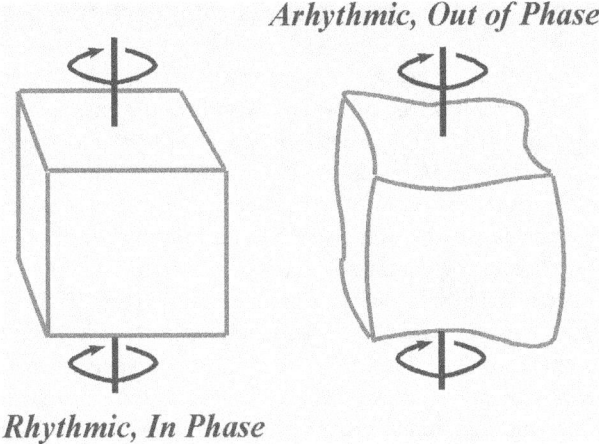

Figure 6.11 Rotating wire cubes. © Stephen E. Robbins.

pass it through the hologram of Figure 6.3, a composite, fuzzed image of both the pyramid-ball and the cup is specified. This too is an optimal specification based on information in the hologram. The rotating, plastically deforming not-a-cube is in fact a vast, extended 4-D structure in time, in this case, again, a structure being *optimally specified* given the strobe/information conditions. This 'optimality', together with retrieving memories from the field and mixing them in the specified perception (another aspect of Bergson's theory), is the basis for the illusions that indirect realism – the 'all is in the head' position – wrongly thinks critical to its position.

Now if we put these invariance laws together relative to some event, say, stirring coffee with a spoon, we get what I term an event *invariance structure*. A partial list of the laws involved:

- A radial flow field defined over the swirling liquid.
- An adiabatic invariant re the spoon, i.e. a ratio of energy of oscillation to frequency of oscillation (Kugler and Turvey 1987).
- An inertial tensor defining the various momenta of the spoon (Turvey and Carello 1995).
- Acoustical invariants.
- Ratios relative to texture gradients and flows for the form, size constancy, even our grasping of the cup (Savelsbergh, Whiting and Bootsma 1991).
- And more...

Given our brain is specifying (or 'specific to', à la Gibson) the coffee cup and the swirling liquid surface 'out there' while we stir the liquid, it is this dynamic invariance structure, with its invariants defined only over time (and all *coordinate* with each other, hence no need for 'binding'), that is *driving the modulation pattern* of the brain as a specifying reconstructive wave.

Bergson – the alternative to the computer model of mind

The brain as a very concrete wave – a reconstructive wave – is very counter to the computer or computational model of mind. Yet there are indications that we are moving to this view of the brain. The neuroscientist authors of the recent book, *The Relativistic Brain* (Cicurel and Nicolelis 2015), argue that recent findings in neuroscience indicate that the neural processes, when taken in larger, group scale, are creating electro-magnetic fields. In other words, the brain is indeed a very concrete device, as concrete as an AC motor generating an electric field of force.

In the computational model, abstract computations are given all the work, and any device is sufficient – from an abacus, to a register machine using beans and shoe boxes, to a Turing machine with an infinite tape – as long as the *computations* can be carried out, the *concrete dynamics* are irrelevant. However, abstract computations cannot account for consciousness – or conscious perception. We need a real, concrete, specific dynamics (Robbins 2014). You don't make an AC motor from rubber bands, toothpicks and shoe boxes – or a modulated reconstructive wave. This is why, as folks like Searle hold but cannot quite explain, 'the biology is important'.

In this, we move to the 'broad computation' Turing recognized in his 'O-machines', where, for example, a protein instantly finds the optimal solution to the otherwise computationally intractable problem of its 3-D configuration, simply by following the laws of physics in its concrete, analog domain. What the computational model sees as abstract 'computations' in the brain should be viewed as constraints – via the invariants – embodied in the constantly modulated form of this very concrete wave that is the brain, a wave specifying an image of the external field.

Time in Bergson's model – the temporal metaphysic

This is where the optical problem also becomes a problem in physics' model of time. This specified image is an image of the *past*. When we see a fly buzzing by, its wings beating 200 times per second, we are seeing a blurred summation of an already long past history of the fly's motion, i.e. we are seeing the fly as a *past* transformation of a (small) portion of the external, holographic field. How is this past-specification possible? Here we must bring in Bergson's model of time.

Underlying current physics is the *classic metaphysic* of space and time. Relativity is simply the logical epitome of this metaphysic. Beneath the metaphysic is an *abstract space*. This space is, in Bergson's terms, 'a principle of infinite divisibility'. When an object, say, our buzzing fly, moves from point A to point B, it is conceived to trace a trajectory or line (a space) (see Figure 6.12). The line consists of a set of points, and each point momentarily passed over by the object is conceived to correspond to an 'instant' of time. Thus time, in this framework, is treated as just another dimension of this abstract space – the dimension of 'instants', each instant corresponding to an instantaneous state or snapshot of the 3-D space. However, when motion is treated merely as a set of points, and as the line is infinitely divisible, then to explain the motion between each pair of static, immobile points, we must insert a new line with its points,

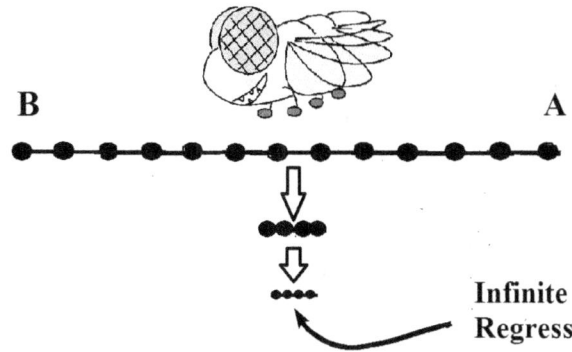

Figure 6.12 Motion treated as a set of static points. © Stephen E. Robbins.

beginning the process over again – ad infinitum, i.e. an infinite regress. This treatment, Bergson argued, is the root cause of Zeno's paradoxes. Achilles, continually halving the distance to the tortoise as he runs along this infinitely divisible line or space, never catches the tortoise. The flying arrow, per Zeno, 'never moves', for at any 'instant' it is at rest at a static point in this space.

For Bergson, motion (thus the transformation of the universal field, namely, time) must be treated as *indivisible*. Achilles moves in indivisible steps; he most certainly catches the tortoise. Motion, Bergson argued, should be conceived as a melody where each 'note' (instant) interpenetrates or permeates the next, each reflecting the entire series, forming an organic whole. From this perspective, the motion of time itself carries an elementary form of *memory*. Appropriating a phrase from William James (1890), we can term it (the true) *primary memory*.

In the classic metaphysic, where time (or the motion of the ever-transforming universal field) consists of a series of discrete 'instants', each instant falls into the past when the next instant (the 'present' instant) arrives. As the 'past' is our symbol of non-existence, the brain, itself only existing for the duration of that instant, is assigned the task of somehow instantly storing each 'present' instant to preserve it. No. This transformation is indivisible; it carries a primary memory. Therefore, the brain, as a reconstructive wave, is perfectly able to optimally specify the past motion of the holographic field – stirring spoons, buzzing flies, falling, twisting leaves, plastically deforming not-quite-cubes – all transforming in an indivisible flow. The brain is not relying on mythical, logically impossible, short term storage areas of static memory (yet to be found) to store these flows (Robbins 2020).

This notion of the indivisibility of motion or time is knocking at physics door. Nottale, in 1996, argued, very bluntly, that space-time is non-differentiable. Differentiation implies infinite divisibility, as when we divide the slope of a triangle, or a motion from point A to point B, into successively smaller sections – ultimately 'taking a limit' to arbitrarily end what is in fact an infinite operation. In the fractal context of Nottale, with the awesome implications of the nature of fractals, every where one looks at the geodesic curves of space-time, ultimately at the most infinitesimal of scales, one finds an inflection point, meaning – the curves cannot be differentiated. Lynds, in 2003, echoing Bergson, implicitly also reinforcing Nottale from a different direction, argued that there can be no static instant underlying any dynamic physical process. There is constant change. No matter how infinitely small the interval examined, there is change. If there were such a truly static instant, the entire universe would be frozen, never to change again. No value then can be fixed with certainty. Every equation of physics is subject to uncertainty. It is an intrinsic tradeoff – uncertainty for constant change

Thus, Bergson argued that there must be *real* motion (MM: 254). Any and all motion cannot become *rest* simply upon perspective, as in the very mistaken current interpretations of relativity (this is another story I must neglect here, though cf. Robbins 2010, 2013). We may not be able to say which objects are in motion, which objects are at rest, but real motion there must be. Stars explode. Trees grow. Mountain ranges arise. We must view the *whole as changing*, he argued, as though a kaleidoscope. Thus, he stated, what we term the 'motions' of separate 'objects' become *changes or transferences of state* within this global transformation or motion (like waves in the sea). It is a transformation with an inherent simultaneity, and it is indivisible – like a melody.

Back then to our buzzing fly, specified as a portion of the past, indivisible transformation of the matter-field: the fly, its wings a-blur, is also a reflection of the *scale of time* imposed upon the field by the dynamics – physical and chemical – of the brain. Here we go to an implication (clearly seen by Bergson, MM: Ch. 4) that the authors of *The Relativistic Brain* did not see. These authors envisioned a constraint on the global processing velocity of the brain, but the value of this constraint can be changed. Introduce a catalyst or set of catalysts into the brain. A catalyst supports an increase in the velocity of chemical flows that otherwise would require a higher input of energy, and thus it can raise the chemical velocities in the neural processes supporting this dynamic 'wave' that is the brain. Raise the chemical velocities (or, shorthand – the *energy state*) to a certain level – the fly is now specified as flapping its wings slowly, like a heron. Increase the velocities yet more – the fly is now motionless, its wings moving not at all. Raise the velocities yet more – we begin to see the fly as the liquid, vibrating, crystalline structure that it is (cf., Robbins and Logan, forthcoming). The holographic field can be 'specified' at an infinity of scales of time.

Note, by the way, this very time-scaling *is qualia*, but find this anywhere in the vast discussion on the hard problem – the problem of time is utterly ungrasped. Note too that all this is integrally related to perception as *virtual action*, i.e. that the brain has selected information to specify from the holographic field on the very basis of action. The heron-like fly must be a veridical specification of how we can *act*. In this case, that we could reach out slowly and grasp the fly by his wing-tip. This is enactivism at its ultimate. It should ultimately be testable.

Time scales and invariance laws

If we can do this in principle, that is, raise the energy state of the brain, then we must assume that nature has allowed for it (note, even increasing temperature increases chemical velocities). This reinforces why it is Gibson's invariance laws that are *required*,

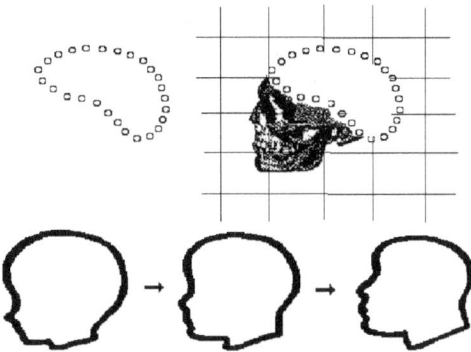

Figure 6.13 The aging of the facial profile – a strain transformation applied to a cardioid. © Stephen E. Robbins.

that are necessary for specification of the external world. Effectively, by changing scales, then, somewhat analogous to relativity, we are changing the 'space-time partition'. The essence of relativity is that it is *only* invariance laws that hold across all partitions, e.g. we have d=vt in one observer's reference system (say, at rest), and d'=vt' in another observer's reference system (in motion).

As an example, take Pittenger and Shaw's (1975) law for the aging transformation of the human head (Figure 6.13). Aging is a very slow transformation in our normal scale of time. The head growth or change is specified by a strain transformation on a cardioid figure placed over the skull and placed upon a coordinate system. Strain stretches the cardioid (and skull) in all directions as though it were on a rubber sheet. Increasing the strain value increases the 'aging'. Were the head transforming rapidly before us – a very fast scale – it is yet this strain law (an invariance law) that would be specifying how to modulate the hand to grasp the rapidly transforming head.

This brings us to another of Bergson's obscure passages: 'Questions relating to subject and object, to their distinction and their union, must be put in terms of time rather than of space' (MM: 77).

Subject, object and time – Bergson's unique panpsychism

We have seen that Bergson viewed the universal field as holographic. We have seen that he held the dynamic transformation of this holographic field as indivisible, where each 'instant' interpenetrates, permeates the next, as the notes of a melody, and where each note (instant) reflects the entire preceding series of notes. There is a deep implication in this. If the state of each point/event in the field reflects the influences of the whole, in fact the *history* of the whole, then, in effect, each 'point' at the null scale of time (i.e. at the infinitely smallest scale) has an elementary awareness of the whole. Bergson called this 'pure perception.' It is as though, stretched across the universal matter-field, is a vast, vibrant 'web' of awareness. It is a highly *coherent* web; the threadlike 'fibres' are taut, a light flick with a finger sends reverberations instantly throughout the whole. This 'web', with its, (a) basic, elemental awareness and, (b) its fundamental, primary memory as the web transforms indivisibly over time, carries the elementary attributes of mind. Yes, this is a unique *panpsychism* – we do not need tiny, 'proto-conscious' particles here, where the consciousness of each is somehow, in some mysterious, inexplicable way, *aggregated* or 'combined' (Chalmers 2016) together to form a larger consciousness (i.e. our experience – our perception of the stirring coffee) such as that of humans, or even of chipmunks. The concept of the brain as a specifying reconstructive wave is the resolution to this 'aggregation' problem.

So, note, this is the *null scale* of time, not the scale of buzzing flies, stirring spoons or falling leaves. Whether the chipmunk's or the human's, the brain – as a reconstructive wave embedded in this field, passing through it – is specifying past portions of the change of this field at a particular scale of time – a buzzing fly or a heron-like fly. The brain is establishing a certain *ratio* relative to the micro-events of the matter-field, in the fly's case, the micro-events making up the body of the fly, its wing beats, its internal processes. If we were to conceive of our body and the fly side by side within the universal field at the null or infinitely small scale of time, we see there is no spatial differentiation between our body and the fly. Both of these 'objects' are simply phases – transferences of state – within the global transformation of this field. However, allow the brain to gradually apply an increasing time scale in its specification: the outlines of the fly begin to emerge, then the shimmering oscillations of his vibrant, organic crystalline structure, then slowly he begins to flap his wings, and then becomes the buzzing being of normal scale. The essential unity of the two within the matter-field – our body and the little fly – is never broken. We arrive at Bergson's principle: *subject is differentiating from object, not in terms of space, but of time*.

To give a respectful bow to our Japanese friends here, let me slightly modify one of those koans of Zen, this one from the master, Bassui. From, 'Who is it that hears?' the koan becomes, 'Who is it that sees?' The basis of the experiential answer of Zen enlightenment is clear: there is no one that sees. What is being specified – at a *given scale of time* – is a time-scale-modified, perspective-based, bodily action-relevant form of the elementary awareness defined throughout the holographic field.

Yes, in such a solution to the hard problem, in such a model of perception, there are profound consequences for the nature of memory, cognition and thought, all of which are waiting to be explored (cf., for a start, Robbins 2002, 2006b, 2009, 2017, 2020) and which begin to form a complete and concrete alternative to the computer model of mind.

So, this is Bergson – the great and unique panpsychist, the original externalist, the radical enactivist, the powerful direct realist – all unrecognized, not only by Searle, but by most of the philosophic world.

References

Bergson, H. (1911), *Matter and Memory*, trans. N.M. Paul and W.S. Palmer, London: George Allen and Unwin, https://archive.org/details/in.ernet.dli.2015.506437/page/n7

Bohm, D. (1980), *Wholeness and the Implicate Order*, London: Routledge and Kegan-Paul.
Caufield, H.J. and S. Lu (1970), *The Applications of Holography*, New York: Wiley and Sons.
Chalmers, D. (2016), 'The Combination Problem for Panpsychism', in G. Brüntrup and L. Jaskolla (eds), *Panpsychism: Contemporary Perspectives*, 1–37, Oxford: Oxford University Press.
Cicurel, R. and M. Nicolelis (2015), *The Relativistic Brain: How it Works and Why it Cannot be Simulated by a Turing Machine*, Durham: Kios Press.
Hardcastle, V.G. (1995), *Locating Consciousness*, Philadelphia: John Benjamins.
James, W. (1890), *Principles of Psychology*, New York: Holt and Co.
Kim, N., M. Turvey and C. Carello (1993), 'Optimal Information About the Severity of Upcoming Contacts', *Journal of Experimental Psychology: Human Perception and Performance*, 19 (1): 179–93.
Kock, W.E. (1969), *Lasers and Holography*, New York: Doubleday-Anchor.
Kugler, P. and M. Turvey (1987), *Information, Natural Law, and the Self-assembly of Rhythmic Movement*, Hillsdale, NJ: Erlbaum.
Lynds, P. (2003), 'Time and Classical and Quantum Mechanics: Indeterminacy Versus Discontinuity', *Foundations of Physics Letters*, 16 (3): 343–55.
Nottale, L. (1996), 'Scale Relativity and Fractal Space-time: Applications to Quantum Physics, Cosmology and Chaotic Systems', *Chaos, Solitons and Fractals*, 7 (4): 877–938.
Pittenger, J.B. and R.E. Shaw (1975), 'Aging Faces as Viscal Elastic Events: Implications for a Theory of Non rigid Shape Perception', *Journal of Experimental Psychology: Human Perception and Performance* 1: 374–82.
Pribram, K. (1971), *Languages of the Brain*, New Jersey: Prentice-Hall.
Robbins, S.E. (2000). 'Bergson, Perception and Gibson', *Journal of Consciousness Studies*, 7 (1): 23–45.
Robbins, S.E. (2002), 'Semantics, Experience and Time', *Cognitive Systems Research*, 3 (3): 301–55.
Robbins, S.E. (2004), 'On Time, Memory and Dynamic Form', *Consciousness and Cognition*, 13 (4): 762–88.
Robbins, S.E. (2006a), 'Bergson and the Holographic Theory', *Phenomenology and the Cognitive Sciences*, 5 (3): 365–94.
Robbins, S.E. (2006b), 'On the Possibility of Direct Memory', in V.W. Fallio (ed.), *New Developments in Consciousness Research*, 1–64, New York: Nova Science.
Robbins, S.E. (2009), 'The COST of Explicit Memory', *Phenomenology and the Cognitive Sciences*, 8 (1): 33–66.
Robbins, S.E. (2010), 'Special Relativity and Perception: The Singular Time of Psychology and Physics', *Journal of Consciousness Exploration and Research*, 1 (4): 500–31.
Robbins, S.E. (2013), *The Mists of Special Relativity: Time, Consciousness and a Deep Illusion in Physics*, Atlanta: CreateSpace.
Robbins, S.E. (2014), *Collapsing the Singularity: Bergson, Gibson and the Mythologies of Artificial Intelligence*, Atlanta: CreateSpace.
Robbins, S.E. (2017), 'Analogical Reminding and the Storage of Experience: The Hofstadter-Sander Paradox', *Phenomenology and the Cognitive Sciences*, 16 (3): 355–85.
Robbins, S.E. (2020), 'Is Experience Stored in the Brain? A Current Model of Memory and the Temporal Metaphysic of Bergson', *Axiomathes*.
Robbins, S.E. and D. Logan (forthcoming), 'LSD and Perception: The Bergson-Gibson Model for Direct Perception and its Biochemical Framework', *Psychology of Consciousness*.

Savelsbergh, G.J.P., H.T. Whiting and R.J. Bootsma (1991), 'Grasping Tau', *Journal of Experimental Psychology: Human Perception and Performance*, 17 (3): 315–22.

Searle, J. (2015), *Seeing Things as They Are: A Theory of Perception*, Oxford: Oxford University Press.

Turvey, M. and C. Carello (1995), 'Dynamic Touch', in W. Epstein and S. Rogers (eds), *Perception of Space and Motion*, 401–90, San Diego: Academic Press.

Weiss, Y., E. Simoncelli and E. Adelson (2002), 'Motion Illusions as Optimal Percepts', *Nature Neuroscience*, 5 (3): 598–604.

7

Bergson and Ecological Psychology: Memory of the Body and of the Universe

Tetsuya Kono
Rikkyo University

1 Introduction

In this chapter, I will focus on Bergson's theory of perception and memory; especially I will try to clarify his famous claim that the brain is not the place for memory. Bergson said that 'memory exists independent of the brain' and 'memory is something other than a function of the brain' (MM: 315).

Bergson has rarely been to referred in the literatures of the philosophy of mind and of cognitive science. Even phenomenologists don't talk so much of Bergson these days. It is well known that Merleau-Ponty criticized Bergson's *Matter and Memory* as committed to an old-fashioned realism (Merleau-Ponty 1978). Merleau-Ponty has also been often referred to and admired in the philosophy of cognitive science. From this, Bergson's theory of perception and memory might have been considered as a 'proto-phenomenology' which was overcome by Merleau-Ponty, who regards the world from the phenomenological standpoint as the object of intentionality rather than simple reality. However, I think that Merleau-Ponty didn't fully understand the radicalism of Bergsonian philosophy. When Bergson said that memory exists independent of the brain, something more radical than the recent theories on the philosophy of the mind, such as 'extended mind', is expressed: memory is not a psychological function, but a metaphysical function; in other words, our memory is another name of the virtuality in the universe (Clark and Chalmers 1998; Chemero 2009; Gallagher 2008; Noë 2009; Rowlands 2011; Thompson 2007; Wilson 2004). All psychological theories about memory must be revised from the point of view of Bergsonian philosophy.

In this chapter, I would like to elucidate what Bergson intended to maintain by referring to the independence of memory from the brain, and to revive Bergsonian theory of memory and perception through the comparison with the ecological psychology of James J. Gibson. Ecology is the biological study of the relationship between the animal and its environment or niche. Gibson adapted ecological thinking into the domain of psychology, especially perception psychology. The most important concept of Gibson's psychology is the 'reciprocity', which affirms that the perception consists of the circular process between the animal/perceiver and the environment. Gibson also affirms that biology and psychology must begin with an environment, a medium, and surfaces (solid, viscous, liquid) that undergo change (Gibson 1979: 16–32).

Some researchers have already pointed out the resemblance between Bergson and Gibson (Ex. Nakamura 1986; Kurenuma 2009; Robbins 2000). Through the comparison with Gibsonian psychology, I will try to show that Bergson is not only a pioneering figure for the post-representationalist theory of mind, but also that Bergsonian theory would provide a deeper understanding of perception and memory than the current philosophy of mind and of cognitive science as well as psychology have. What I will try to do, ultimately, through this comparative study is to establish an alternative psycho-metaphysical theory of memory that can take the place of the prevailing, but wrong theory of memory in psychology, namely the theory that believes that memory is stored in and retrieved from the brain. Both Bergson and Gibson refused the storage-retrieve theory of memory.

2 Bergsonian theory of perception and J.J. Gibson

When Bergson said that 'all we sense are images' and 'the matter is for us the totality of images' (MM: 6), his idea of 'image' must be interpreted as direct or 'naïve' realism. Direct realism is defined as an ontological and epistemological view which claims that the world is what we perceive. Direct realists refuse both materialism and idealism. Bergson criticizes materialism because it presupposes hidden powers that are able to produce representations in us. Distinctions of matter and representation and that of primary qualities and secondary qualities surely provoke unsolvable 'hard problem' or 'explanation gap'. For Bergson, matter is only images; there is no hidden power in matter.

Bergson also criticizes idealism insofar as it attempts to reduce matter to the representation we have of it. Image differs from representation but it does not differ in nature from representation. An image may be without being perceived; it may be present without being represented. Representations are some parts of all images in the universe. The relation between the representation and the image, in other words, the phenomenon and the thing is not that of appearance to reality, but merely that of the part to the whole. Bergson said:

> I call matter the aggregate of images, and perception of matter these same images referred to the eventual action of one particular image, my body.
>
> MM: 8

> My perception can, then, only be some part of these objects themselves; it is in them rather than they in it.... Perception, therefore, consists in detaching, from the totality of objects, the possible action of my body upon them. Perception appears, then, as only a choice. It creates nothing; its office, on the contrary, is to eliminate from the totality of images all those on which I can have no hold, and

then, from each of those which I retain, all that does not concern the needs of the image which I call my body.

MM: 304

Thus, perception adds nothing new to the image. Rather, it subtracts from it; perception is a selection. Perceptual selection occurs because of necessities or utility based in our bodies. So, representation is a diminution of the image. Representation results from the suppression of what has no interest for bodily functions and the conservation only of what does interest bodily functions. The conscious perception of a living being therefore exhibits a 'necessary poverty' (MM: 38).

For Bergson, consciousness lies in choice, or discernment. What Bergson maintained about the brain function anticipates development of contemporary selectionist theories of the brain such as Young (1979), Edelman (1987), Changeux and Dehaene (1989) and others. Contemporary selectionists expanded Darwinian selective systems into the fields of neurobiology and neurophysiology affirming that the neural system in each individual animal operates as a selective system, but operating by different mechanisms from natural selection in evolution, but I don't argue this point any further in this chapter.

For Bergson, perceptual world is already there, and perception is a 'slicing up' or a 'selection' of some part of it, i.e. pay attention to some part of it. This theory is very close to James J. Gibson's theory of direct perception and specification of information. Gibson (1966, 1979) is well known as a theorist of direct perception. To say that perception is direct means that the real world is as we perceive. The world is not a simple mass of atomistic sense data without structure which needs to be enriched by internal cognitive processes (memory, inference, computation, etc.) of the perceiver. Gibson (1966, 1979) strongly criticized the indirect perception theory for presupposing that the stimulus information available to the senses is ambiguous and needs to be enriched.

Instead, Gibson proposed that perceptual information is specific to environmental properties; information relates one-to-one to it (Withagen and Kamp 2010). He argued that such specifying information resides in the ambient arrays, the energy arrays that surround us. Gibson claims that perception is specifying invariants from the flux or variants of stimulation. Our perceiving body, being exposed to the flux of stimulation, tries to specify stable and invariant features of the environment. Perception is an act. Perception is thus the act of picking up or detecting some invariants from the changing flow of stimulation. The 'image' in Bergsonian philosophy is the ecological 'reality' in Gibsonian psychology.

3 Two types of memories of Bergson

It is well known that Bergson distinguished two types of memory: habit-memory and pure memory. Habit-memory is obtaining certain automatic behaviour, i.e. the acquisition of sensori-motor schemas by the means of repetition. Habit-memory is the procedural stored mechanisms inscribed within the brain, aligned with perception, for unrolling an intentional action. It replays and repeats past action, not strictly recognized as representing the past, but using it for the purpose of present action. Bergson thinks that this mechanism is a function of brain.

> [I]n fact, it is towards action that memory and perception are turned; it is action that the body prepares.... We note that its primary function is to evoke all those past perceptions which are analogous to the present perception, to recall to us what preceded and followed them, and so to suggest to us that decision which is the most useful.
>
> MM: 30

Our representation of matter is the 'measure of our possible action upon bodies' (MM: 30). In this view, regarding perceptual objects as those of possible actions, Bergson resembles Gibson. Gibson coined the noun affordance. Gibson defined affordance as follows: 'The affordances of the environment are what it offers the animal, what it provides or furnishes, either for good or ill.... I mean by it something that refers to both the environment and the animal in a way that no existing term does. It implies the complementarity of the animal and the environment' (Gibson 1979: 127). He explains that the offering of the affordances are sets of promises (positive affordances) and threats (negative affordances) that characterize items in the environment relative to animals.

For example, the support by the surface is an example of positive affordances. If a terrestrial surface is nearly horizontal, nearly flat, and sufficiently extended, and its substance is rigid enough, the surface affords support to an animal. 'It is stand-on-able, permitting an upright posture for quadrupeds and bipeds. It is therefore walk-on-able and run-over-able' (Gibson 1979: 127). Thus, affordances are expressed as 'verb-able'. A lake is 'sink-into-able', bread is 'eatable', burning fire is 'cook-with-able' but at the same time 'getting-burned-able' and a knife is 'cut-with-able' but at the same time 'being-cut-able' or 'being-hurt-able'.

Thus, affordances are potentialities or dispositional properties of the environment that yield an ecological event in which an animal itself is caught up either positively or negatively by the presence or the intervention of an animal to that environment. The presence of an animal is itself a set of activities such as breathing, heating, weighing, pushing and so on. Accordingly, affordances express the circular functional processes between an animal and its environment. This is why affordances are regarded as ecological value or meaning of the environment. Because affordances are the potentialities of the environment, information about affordances involves the information about the future events in the environment and the future of the animal itself. Affordances are necessary conditions for animal actions; because the perception of affordances is that of the future, it rends possible intentional, purposive behaviour.

We shall continue explaining Bergsonian concept of memory. In contrast with habit-memory, pure memory, true memory, registers the past in the form of image-remembrance, representing the past, recognized as such. It corresponds to the episodic memory in our psychology. Pure memory permits the acknowledgment that the lesson has been learned in the past, cannot be repeated, and is not internal to the body. Pure memory begins with the metaphysical, and no longer merely psychological bearing.

Bergson explained the relation between memory and perception using the figure below. The image of the cone is constructed with a plane and an inverted cone whose summit is inserted into the plane. The plane, 'plane P', as Bergson calls it, is the 'plane of my actual representation of the universe'. The cone 'SAB', of course, is supposed to symbolize memory, specifically, the true memory or regressive memory. At the cone's base, 'AB', we have unconscious memories, the oldest surviving memories, which come forward spontaneously, for example, in dreams.

There is not a mere difference of degree between recollection and perception but a radical difference of kind. Memories are descending down the cone from the past to the present perception and action. True memory in Bergson is progressive. This progressive movement of memory as a whole takes place between the extremes of the base of 'pure memory' and the plane where action takes place.

> The truth is that memory does not consist in a regression from the present to the past, but, on the contrary, in a progress from the past to the present. It is in the past that we place ourselves at a stroke. We start from a virtual state which we lead onwards, step by step, through a series of different planes of consciousness, up to the goal where it is materialized in an actual perception; that is to say, up to the point where it becomes a present, active state; in fine, up to that extreme plane of our consciousness against which our body stands out. In this virtual state pure memory consists.
>
> MM: 319

Our past ... is that which acts no longer but which might act, and will act by inserting itself into a present sensation of which it borrows the vitality.

MM: 320

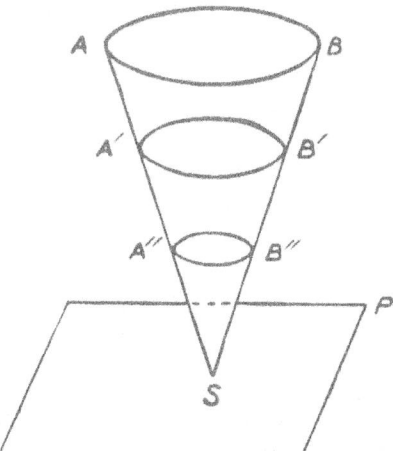

Figure 7.1 Bergson's inverted cone. Henri Bergson. Public domain.

For Bergson, the past does not disappear; it is not gone. It still exists as virtuality that can be materialized in an actual perception. Our present is the actual state of our body, while pure memory is the virtuality. The present is conscious and the past is unconscious.

4 Depth perception for Gibson

James J. Gibson's ecological psychology provides us a more important suggestion about temporal structure of our experience. Newtonian physics and mathematics begin with an empty three-dimensional space, the dimension of time and bodies that move. However, for Gibson there is no such thing as the perception of space but only the perception of environmental layout; there is no such thing as perception of time but only of event.

The notion of events is also basic to Gibson's theory of perception (Gibson 1982: 203–16). 'We should begin thinking of events as the primary realities and of time as an abstraction from them. . . . Events are perceived, but time is not' (Gibson 1979: 100). An event is a change such as motion, action, transformation, transfiguration and so on, which has a beginning and an end, it can be repeated and a sequence of events exists. In a sequence of events in the natural environment, there is no fixed frequency, because beginnings and endings depend on the duration of event we select for attention. The number of discrete events cannot be counted. Therefore, we cannot have a numerical 'span' of apprehension or memory.

According to Gibson, an event is directly perceived. Perception of sequences does not depend on memory in the sense of storage and retrieval of traces. In fact, psychological theories about memory storage and its retrieval are becoming more and more complex. This is the result of the acceptance of the old-fashioned framework of the modern (not contemporary) physics that space and time are distinct dimensions and time consists of distinct moments, so memory is needed to connect separated 'nows'. Contrary to this prevailing theory, Gibson maintains that the perception of events does not depend upon recall or recollection. He criticizes the 'stream' theory of consciousness or experience:

> The stream of consciousness . . . exhibits the traveling moment of present time, the feeling of 'now', and this divides the stream into a past extending in one direction and a future extending to the other. But the stream of stimulation has no such unique moment. Neither does the flow of external events.
>
> Gibson 1982: 173

To perceive the environment, it is neither necessary to refer to the past experience, nor to compare the present experience with the past experience. Surely, there is perceptual learning: differences are noticed that were previously not noticed; features become distinctive that were formerly vague. It is true that past experience improves perceiving skill, educates the way of attention, and makes perceptual system more sensitive. This is the function of habit-memory in the Bergsonian sense. The state of perceptual system is altered when it is attuned to information of a certain sort. However, 'this altered state needs not to be thought of as depending on a memory, an image, an engram, or a trace in the brain' (Gibson 1979: 254).

For ecological psychology, we *perceive* the phenomenal persistence. It means the perceiving of surfaces that is continuing to exist in the future and having existed in the past as well as in the present. The layout of the environment includes unprojected, namely hidden surface at a point of observation as well as projected surface, but the observers perceive the layout not just the projected surface. Things are seen in the round and one thing is seen in front of another. How can this be? This is the problem of depth. Gibson affirms that 'what we see is not depth as such but one thing behind another' (Gibson 1979: 77).

Perception of depth is not to see surfaces that are unseen; it is of course a paradox. To suggest that we perceive depth means that we can perceive the surfaces that are temporarily out of sight. The important fact is that surfaces come into sight and go out of sight as the observer moves. The situation that the surface goes out of sight is very different from the situation that the surface goes out of existence. Depth perception is the perception of surface-being-revealed and of surface-being-concealed. i.e. going-behind and coming-from-behind. It is the perception of depth. Gibson wrote an extremely important phrase as follows:

> The two [the past and the future] are united in the fact of occulting edges. At an occulting edge the present *hides* the past and also hides the future. During locomotion what is being concealed 'goes into the past' and what is being revealed 'comes from the future.' But actually of course one simply apprehends the whole environment.
>
> Gibson 1982: 396

The past and the future are hidden by the present. Future is the hidden dimension of the world that is concealed by the manifested dimension, the present, of the world. The characteristic of future, 'not yet', is not nothingness that is the negation of existence. We cannot perceive unknown, indefinite matters in the far future, but can only intellectually estimate and calculate them. On the other hand, we perceive possible changes that are in the process of happening and will be realized in the near future. The perception of affordances is such perception. When we are looking at a moving car, we are looking at the car in certain duration. The track that the car is coming from over there to here is the realized potentiality of the car locomotion; we can say that the car is passing in front of me and will go far away very soon is the realizing and will-be-realized potentiality. This is neither expectancy nor estimation, but perception. It is not an intellectual operation, but the perception of the potential embedded in the present situation. We see that the world is continuing and the events happened in front of me will be involved in later events and will develop further. We can perceive the potentiality because potentiality is a real aspect of the world.

The future is not the transparent extension of the line from the past to the present. Time is not a line for Gibson or for Bergson. Actually, if time is a flux or stream like a river, what has passed, or what has gone? Everything? Have natural laws gone too? Such spatial metaphor of time implicitly presupposes more profound reality of temporality, i.e. the potentiality of the world for becoming further. The perception of the future is 'foresight' of the change and the movement (Gibson 1982: 396). According to Gibson, we perceive 'time' in the differentiation of the variable and the invariable and in the change as actualization of the potential. Future is the change as differentiation from the present.

Therefore, future is depth for Gibson as well as for Bergson. Depth means that something is hidden. A hidden thing is the potential of the world; it is revealed under certain conditions. The perception of the future is the perception of the potential embedded in the present situation.

5 The ontological status of memory in the universe

We shall return to Bergson. Where does memory exist if it is not stored in the brain? The relation between perception and memory is like that of an object and its image in mirror:

> [I]t [object] is pregnant with possible actions; it is actual. The image [in mirror] is *virtual*, and thought it resembles the object, it is incapable of doing what the object does. Our actual existence, then, whilst it is unrolled in time, duplicates itself all along with a virtual existence, a mirror-image. Every moment of our life present two aspects, it is actual and virtual, perception on the one side and memory on the other.
>
> ME: 165

According to Bergson, our existence is divided into the actual and the virtual, conscious and unconscious at the same time when it is posed. The formation of memory is never posterior to that of perception but they are simultaneous. Pure memory corresponds to the episodic memory; an episode is an event. As each of the Leibnizian monads is a mirror of the universe, each episode represents the conditions of the whole universe.

> Only if when we consider any other given place in the universe we can regard the action of all matter as passing through it without resistance and without loss, and the photograph of the whole as translucent: here there is wanting behind the plate the black screen on which the image could be shown. Our zones of indetermination [= our living body] play in some sort the part of the screen. They add nothing to what is there that the real action passes through, the virtual action remains.
>
> MM: 31–2

Pure memory starts from a virtual state up to the goal where it is materialized in an actual perception, i.e. up to the point where it becomes a present, active state. The move from pure memory to perception, the virtual to the actual, is the comprehensive process of events. A past event advances by integrating or englobing other more recent past events towards the present in forming a context, an episode. Memory for Bergson is not embodied inside the human body, but it the virtuality of the world which can be actualized, in organizing itself as a nested event by the call from a living perceiving body. The virtual is existing in essence or in effect though not in actual fact. This is the process of the action of the universe.

For Bergson, memory is no longer a psychological faculty but a metaphysical process of the universe. We have to understand that pure-memory is no more than memory in the ordinary sense of the word, but the context making process through integrating past events. Memory is the principle of the becoming of the world. The actual action of human body calls this inclusive process of the virtuality of the universe and surfs on that process in becoming a part of the wave. Thus, memory is the comprehensive process of events of the world. A human body is part of an enormous body of the universe in process.

6 Concluding remarks

According to Gibson, when we perceive 'the present' of the environment, we perceive simultaneously persistence from the past and the potentiality of change in the future. We perceive affordances of the environment. Affordances are potentialities of the environment. Perception of the future is the perception of the potential in the world. The potential is the possibility for change that is inscribed in the persisting, present environment.

For Bergson, pure memory is the comprehensive process of becoming. It is the virtuality of the world that can be actualized by the call of a living perceiving body. The potentiality or virtuality of the world is the key concept both for Bergson and Gibson.

This coincidence must not be an accident, because the mentor of Gibson was E.B. Holt, a student of William James. Holt connected James and Gibson, and inspired Gibson to be a radical empiricist. The ontology of Gibson is that of pragmatism. James and Gibson shared the worldview that sees the universe as a becoming process. It would be needless to explain the relation between William James and Bergson.

It is important to notice that both Bergson and Gibson did not think that time is not a linear, stream-like development. The image of time as stream or flow is a misleading metaphor. For them, what exists in the world are events with divers paces, duration and frequency. An event becomes an event when its initial effect finds its goal. An open event without the end still remains in the state of potentiality. An actual bodily action calls the potential and makes it an event. This is a context making process. That is why pure memory does not exist inside the brain. The relation between the present and memory, the actual and the virtual is that of part and whole. The present, the past and the future are not three points on a line of time, but they consist in the differentiation between the actual and the virtual. Both the future and the past belong to the same virtuality of the universe. The present makes the direction from the past to the future by differentiation.

References

Bergson, H. (1911), *Matter and Memory*, reprint (2004) of 1912 MacMillan edition, trans. N.M. Paul and W.S. Palmer, New York: Dover Publications.

Bergson, H. (1920), *Mind-Energy*, trans. H. Wildon Carr, New York: Henry Holt and Co, https://archive.org/details/mindenergylectur00berguoft/page/n6

Changeux, J.-P., and S. Dehaene (1989), 'Neuronal Models of Cognitive Functions', *Cognition*, 33: 63–109.

Clark, A., and D. Chalmers (1998), 'The Extended Mind', *Analysis*, 58 (1): 7–19.

Edelman, G.M. (1987), *Neural Darwinism*, New York: Basic Books.

Gallagher, S. (2008), *Brainstorming: Views and Interviews on the Mind*, Exeter: Imprint Academic.

Gibson, J.J. (1966), *The Senses Considered as Perceptual Systems*, Boston: Houghton Mifflin.

Gibson, J.J. (1979), *Ecological Approach to Visual Perception*, Boston: Houghton Mifflin.

Gibson, J.J. (1982), *Reasons for Realism: Selected Essays of James J. Gibson*, E. Reed and R. Jones (eds), London: Routledge.

Kurenuma, N. (2009), 'Mondai no Shingi to Jituzai no Kubun: Gibson to Bergson no Hoho' (The division between true and false and real in question: The Method of Gibson and Bergson), *Shiso*, 1028: 171–92.

Merleau-Ponty, M. (1978), *L'union de l'ame et du corps chez Malebranche, Biran, et Bergson*, 1978, Paris: J. Vrin.

Nakamura, M. (1986), 'Chikaku to Koudo: Bergson Tetsugaku ni okeru Gibson Shinrigaku no Yokose' (Perception and Behavior: The Validity of Gibsonian Psychology for Bergsonian Philosophy), *Nenpo-Ningenkagaku*, 7: 69–82.

Noë, A. (2009), *Out of our Heads: Why You Are Not Your Brain, and Other Lessons from the Biology of Consciousness*, New York: Hill and Wang.

Robbins, S.E. (2000), 'Bergson, Perception and Gibson', *Journal of Consciousness Studies*, 7 (5): 23–45.

Rowlands, M. (2011), *The New Science of the Mind: From Extended Mind to Embodied Phenomenology*, Cambridge, MA: MIT Press.

Thompson, E. (2007), *Mind in Life: Biology, Phenomenology, and the Sciences of Mind*, Cambridge, MA: Belknap Press of Harvard University Press.

Wilson, R.A. (2004), *Boundaries of the Mind: The Individual in the Fragile Sciences: Cognition*, Cambridge: Cambridge University Press.

Withagen, R. and J. van der Kamp. (2010), 'Towards a New Ecological Conception of Perceptual Information: Lessons from a Developmental Systems Perspective', *Human Movement Science*, 29: 149–63.

Young, J.Z. (1979), 'Learning as a Process of Selection and Amplification', *Journal of the Royal Society of Medicine*, 72: 801–14.

8

Defining Philosophy and Cognitive Psychology: Bergson and Bruner

Sébastien Miravète

Cognitive psychology and philosophy are currently two important components of sciences of the mind. How can they be differentiated? How can their approach be characterized? What is cognitive psychology? What is philosophy?

The cognitivist Jerome Bruner and the philosopher Henri Bergson each offered a famous theory of human perception. Therefore, it is impossible to differentiate cognitive psychology and philosophy by their subject matter. They could obviously study the same topic: human perception. They could only be differentiated by their methodological approach. So, the main objective of this work will be to methodologically define cognitive psychology and philosophy by analysing Bruner's and Bergson's theories of perception.

Bruner's cognitive theory of perception and the definition of cognitive psychology

In a famous study, Bruner and Minturn (1955) present an ambiguous figure to participants: the symbol '13'. This symbol is ambiguous because it can be perceived as the letter 'B' or the number '13'. Bruner and Minturn show that participants perceive this symbol as a letter if it is presented with letters ('A 13 C') and as a number if it is presented with numbers ('12 13 14'). This result is very important. It exemplifies that experimental psychology can study the psychological mechanisms of perception. Indeed, this result proves that participants have two internal representations. With a letter representation, participants can identify that 'A' and 'C' are letters. That is why the ambiguous symbol is perceived as 'B'. The context (letters 'A' and 'C') triggers the activation of letter representation and associates the ambiguous symbol '13' to the letter 'B'. Without this letter representation, it is impossible to explain why participants perceive the letter 'B' and not the number '13'.

Therefore, humans could acquire some knowledge (e.g., letter and number representation) and use this knowledge to spontaneously recognize some perceived objects (e.g., the letters 'A' and 'C', the symbol '13',...). So, any act of perception requires sensory-motor organs (e.g. eyes) and knowledge. Perception does not just consist of receiving sensory information. It also consists of *processing* this sensory *information* by using previous knowledge to recognize, evaluate, etc. this sensory information. Perception is a cognitive system, that is, a system of sensory information processing.[1]

In Bruner's theory of perception, there is another important idea. Ambiguous figures have been used in cognitive psychology since 1935 (Wright 1992). In many of these studies, psychologists trigger representations before presenting an ambiguous figure. For example, they say, 'This is a young woman' or 'This is an old woman' (Leeper 1935), and they present an ambiguous figure with a young and an old woman. Levine, Chen and Murphy (1942) expose hungry people to ambiguous figures with food and without food. In these two situations, the cognitive system anticipates perception of objects. Indeed, someone tells the cognitive system that it is going to see an old woman, so it expects to see an old woman. It is hungry, so it perceives food rather than anything else. With its internal representations (Bruner calls them 'categories'), the cognitive system could already be attracted by some things before actually seeing them or not seeing them. In a previous experiment, without ambiguous figures, Bruner and Goodman (1947) discovered that poor children underestimate the size of coins, while rich children have a good estimation of their size. In summary, these experimental results prove that an act of perception is oriented by different kinds of representations: academic knowledge (e.g. letter representation), internal motivation (e.g. hunger), external suggestion (e.g. 'This is a woman'), social representation (e.g. representation of the size of coins by poor and rich children), etc.

This is why cognitive psychology goes beyond the old reductive conceptions of experimental psychology. Cognitive psychology is definitely not Gestalt theory, psychophysics, or behaviourism. Gestalt theory and psychophysics just explain the production of sensory information by sensory factors.[2] They do not study the treatment of this information. They think that this treatment is posterior to the act of perception. In reality, perception is already an act of knowledge. Perception is already cognitive (Fraisse 1949). Behaviourism considers that the experimental psychology cannot explain why a stimulus produces a response.[3] Experimental psychology can just verify that this stimulus triggers this response. The work of Gestaltists, psychophysicists and cognitivists proves otherwise. Response could be experimentally explained by relations between external factors (stimuli) and internal factors (sensory and cognitive factors).

In summary, Bruner's theory of perception exemplifies that cognitive psychology consists of experimentally studying the role of internal representations to explain the behaviour of humans and some animal species. Therefore, cognitive psychology could be defined by its specific methodological approach: (1) using experimental protocols and (2) explaining behaviours by cognitive factors (motivation, reasoning, learning, folk and academic knowledge, etc.).

Bergsonian theory of perception and the definition of philosophy

Bergson describes the mechanism of perception as follows. A physical movement produced by an external object (e.g. light rays reflected by an object) comes into contact with sensory organs (e.g. eyes). This movement extends towards sensory centres (e.g. occipital lobe) located in the brain. Then, some sub-movements of this internal movement located in sensory centres could be repeated. This repetition is freely triggered by imaginary centres and is used to emphasize some characteristics at the expense of others (MM: 165). Thereby, some emphasized characteristics trigger some motor responses (grabbing, throwing, talking, writing, ...) located in motor centres. Other characteristics trigger other motor responses. In brief, imaginary centres enable to freely envisage several motor responses.

A motor response is not just a set of neural mechanisms used to activate and control motor organs (e.g. arms). For Bergson, it is a 'motor schema' of a perceived object. It is a set of neural networks used to decompose and recompose a perceived object. In this manner, the brain knows, in advance, where elements of perceived objects are situated. It knows the position of elements in relation to each other. In short, it knows the 'internal structure'(*structure interne*) (MM: 138) of a perceived object. Therefore, in the Bergsonian theory of perception, the brain has a procedural[4] representation[5] of a perceived object.

In addition, these procedural representations of perceived objects could be freely activated by imaginary centres. Thereby, an imaginary centre is used to question sensory data: what is the correct motor response? What is the structure of this perceived object? etc. Imaginary centres could highlight some details of an object by repeating some sensory data and analysing sensory data with a motor schema. In other words, the imaginary centre could process sensory information. Perception, imagination, and motor centres are like working memory. This is sufficient to conclude that, for Bergson, the nervous system is a cognitive system. In the Bergsonian perspective, perception is not like a reflex. It is already an act of knowledge.

In 1896, the Bergsonian concept of the imaginary centre was a new idea. In 1874, Wundt hypothesized the existence of a centre of attention (Forest 2012: 30). However, in the Bergsonian approach, the imaginary centre is not only used to emphasize some details of the perception by repeating sub-movements in the sensory centre. It is not just a centre of attention. It is also used to process sensory data with procedural knowledge (motor schema). The sensory-motor-imaginary centre is a cognitive model of perception. Therefore, the Bergsonian theory of perception contains the first or one of the first cognitive models of perception in psychology.

Forest (2012) shows the role of pure memories in the processing of perceptual information. Our analysis completes his perspective by showing the role of motor schemas. Our interpretation shows that the brain is already a cognitive system for Bergson, *even without pure memories.*

Bergson justifies the existence of the imaginary centre as follows. Bain and Ribot suppose that memories of the perceived object are contained in sensory centres. Shaw and Luciani demonstrate that these same memories could be inaccessible even if sensory centres are intact. Thus, the memories are (Bain and Ribot) and are not (Shaw and Luciani) in sensory centres. According to Bergson, the existence of the imaginary centre could be supposed to resolve this empirical contradiction[6] (MM: 166). When the relation between sensory and motor centres is damaged, perception becomes very confused or impossible (e.g. partial or total blindness). For example, without motor schema, the brain is not able to see the depth, to dissociate elements of a perceived object, etc. When the relation between sensory and imaginary centres is damaged, perception is not confused but impoverished. In this situation, some motor schemas cannot be activated by an imaginary centre. Without some motor schemas, some pure memories (declarative memories: semantic and episodic) and some procedural abilities (procedural representations) become inaccessible. Indeed, with the Bergsonian theory of recognition, motor schemas are able to represent objects with procedural representations but also filter and select pure memories. In this manner, some knowledge (procedural, semantic, episodic) cannot be used when some motor schemas disappear. That is why, for Bergson, Wilbrand's patient can clearly see a city without knowing that this city is his city (MM: 108).

This conciliation of psychological data (Bain and Ribot) and pathological data (Shaw and Luciani) forces Bergson to suppose that pure memories are not located in the brain. Indeed, Bergson takes seriously Bain and Ribot's thesis: if memories of a perceived object are in the brain, they have to be located in sensory centres; they cannot be elsewhere (MM: 160). That is why Bergson does not aim to put memories in another place and concludes that memories cannot be in the brain if they cannot be in sensory centres. Pathological data (e.g. Wildrand's patient) proves for Bergson that memories could not be in sensory centres. Therefore, memories are not in the brain.

This reasoning is interesting. Note that Bergson's arguments are only empirical and scientific. Analysing psychological and pathological data suggests the existence of imaginary centres, motor schemas, and declarative memories not located in the brain (pure memories). That is why perception is not a full cognitive system (it lacks declarative memories) in Bergson's perspective. In summary, scientific data justify what the Bergsonian theory of perception contains (sensory, motor, imaginary centres, motor schemas) and what this theory does not contain (pure memories). So, for the moment, it is impossible to distinguish philosophy and scientific psychology. The Bergsonian theory of perception is just another old scientific psychological theory of perception.

However, this theory is not only a theory of the nervous system. It is also a theory of knowledge. In the Kantian theory of knowledge, conditions of possibility of perception prevent seeing the real world as it is. These conditions allow to perceive objects, but they completely distort them. That is why it is impossible to perceive and acknowledge the real world. In *Matter and Memory*, Bergson does not want to isolate the subject of perception from the real world. He rejects the Kantian theory of knowledge. He wants to think of the relation between subject and object of perception otherwise.

Bergson assumes that the real world is a set of 'images'. Consciousness is not like a light projector. It does not illuminate real objects. In other words, it does not make them conscious. They are already conscious images. They are already bright. Consciousness is like a photographic film. It does not bring light to real objects. It retains a part of the light. It retains some conscious images. It does not insert some brightness into real objects. It does not change the content of real objects. It does not distort real objects. Of course, all images are not retained, perceived, but they are not deformed. In this way, the Bergsonian theory of knowledge does not isolate the subject of perception from the real world (Montebello 2007: 72-4). We can perceive real objects as they are. Sensory organs select some images. The imaginary centre repeats a part of these selected images. The motor schema describes the organization of an object's elements without deforming the content of these elements. In brief, consciousness, sensory organs and centres of perception (sensory, imaginary, motor) do not alter the content of a perceived object. The Bergsonian concept of images and the Bergsonian approach to the nervous system avoid an idealistic conception of the conditions of possibility of perception: Against Kant, Bergson imagines conditions of the conscious perception that do not isolate the subject of the perception from the real object (Montebello 2003). In this perspective, it becomes possible for sciences and philosophy to acknowledge the real world (realistic metaphysics).

Note that this new perspective is elaborated without scientific empirical data. First, Bergson analyses the semantic content of the terms 'become conscious', 'be not conscious', 'be conscious', 'be a real object', 'conditions of possibility', 'conscious images', etc. Then, he proposes new definitions of these terms. These new definitions enable us to construct a non-idealistic theory of knowledge. Ultimately, semantic analysis is used to resolve this old philosophical problem: is it possible to think that humans can know the real world?

However, the Bergsonian theory of perception is not only a theory of the nervous system and a theory of knowledge. It is also a theory of being (an ontology). For Bergson, the physical matter is the succession of equal time intervals, that is, a succession of time units (Fujita 2007), a 'duration'. Conscious perception is also a succession of equal time intervals, but these equal time intervals are longer. Time units of matter are shorter than time units of conscious perception. That is why Bergson explains that some durations can be faster or slower than others (MM: 274). In Bergson's philosophy, the length of time units defines the 'speed' of time. As in music, when the length of time unit change, the rhythm of time's flow increases or decreases.

There is another difference between these two durations (matter and perception). A material time interval does not contain smaller time intervals. It is the smallest of all time intervals. On the contrary, a perceptive time interval contains smaller time intervals. One perceptive time unit comprises a large number of material time units, but these material time units are perceived confusedly and not distinctly. For example, in one second, red light makes 400 trillion successive vibrations. According to the psychophysicist Exner (MM: 272-3), two successive vibrations must be separated by one time interval of two milliseconds. Otherwise, consciousness does not see two successive vibrations, but one single vibration. Therefore, it would take 25,000 years to distinctly perceive each vibration. However, each present moment, that is, each human perceptive time unit, cannot last 25,000 years. So, it is impossible for human consciousness to see material time units. Material time units are inevitably embedded into each other by consciousness. The length of human time units is too short. The rhythm of human perception is too fast. We cannot see the vibrations of red-light colour. We only see the result of their mixture, the product of their 'interpenetration'.

This ontological description of matter and perception obviously uses the numerical concept of duration: Bergson compares the lengths of time units and performs arithmetical operations (2 milliseconds x [400 trillion − 1]≈25,000 years). Is it a mathematical description of being? The Bergsonian duration is a number of time units, but time units in mathematics and time units in Bergsonian philosophy are not the same (Miravète 2012). In mathematics, time units are simultaneous. They are not successive (on a mathematical timeline, the 'left/right' relation symbolises the 'before/after' relation).

In 1854, the French philosopher Charles Renouvier (Fedi 1999) noted that it is impossible to imagine a succession of time units. For instance, I want to imagine two successive time units. So, I visualize the first, and I visualize the second. In this manner, the first and the second coexist in the same representation. Indeed, I have to remember the first when I think about the second. Otherwise, the first disappears when the second appears, and I never have two units in my representation, only one. In other words, to add time units, to construct a number of time units, to build a duration, we have to position time units side by side. This spatial organization of time units neutralizes their succession. Renouvier concludes that 'duration', that is, real time, cannot be a number of time units.

Bergson knows Renouvier's argument (Miravète 2017), but he continues to think that duration could be a number of time units. To create a new concept of duration, a new concept of number of time units, Bergson analyses semantically the ordinary concept of number in the second chapter of *Time and Free Will*. He notes that units of ordinary numbers are identical. So, units need to be differentiated in space: they are different, because they do not have the same position. It is the same problem with mathematical numbers. For Bergson, the elements of mathematical numbers are not the intervals, but the limits of these intervals. They are geometrical points. They are instants of zero duration. Thus, they are perfectly identical. Therefore, they must be differentiated by their localization in an abstract space. That is why ordinary or mathematical time units are simultaneous for Bergson.

In summary, Bergson understands that time units do not have to be perfectly identical. Otherwise, they must be differentiated in space. Evidently, Bergson cannot use ordinary language to differentiate time units. For instance, when I say 'This man is a soldier and this other man is a soldier', there is no difference between these two men. They are identical because I am only interested in their resemblance ('They are soldiers').Then, I could use other ideas to differentiate these two soldiers. The first is 'big', the second is 'small', etc. However, how can the idea of 'soldier' and the idea of 'big' be differentiated? I could use other ideas to differentiate these two, but then I need new ideas to differentiate these used to differentiate the idea of soldier and big. In brief, I always need ideas to differentiate ideas, and so on. That is why, in the end, when I will exhaust all the words of my language, I will be forced to differentiate persons or objects in space (by their form, their position, etc.). Formal (mathematics, logic) or ordinary language always ends up differentiating objects in space for Bergson. These languages cannot think succession and time.

Bergson uses a sensitive experiment to rethink the old concept of resemblance. His favourite example is that of colour. Two colour shades are never perfectly identical. Two shades of red are similar because they are not blue, but each shade of red is singular: two reds are never perfectly identical. In a sensitive experiment, Bergson finds the model of 'qualitative' resemblance. With this new concept of resemblance, Bergson can redefine time units. Time units in the Bergsonian duration are not perfectly identical. Each time unit is singular. Each time unit is a sensitive 'quality'. Each moment has its temporal flavour.[7] So, it is unnecessary to differentiate time units into space. They differ already. In this sensitive conception, time units could coexist in the same representation and maintain their temporal flavour. Thus, they could coexist and be successive, like notes in a musical melody or sounds of a bell (TFW: 86).

Qualitative time units could coexist without having to be simultaneous. Bergson invents the difference between the coexistence and the simultaneity and exceeds the conclusion of Renouvier. We could conceptualize in philosophy a number of qualitative time units. This number is special and not 'spatial'. In addition, its units could occupy the same interval. They could have the same localization because they do not need to be differentiated into space. Qualitative time units could be literally embodied in each other. That is why one perceptive time unit could contain 400 trillion material time units.

Ontological description of the numerical dimension (length of time units, etc.) of matter and perception is not mathematical. Bergson analyses semantically the mathematical concept of duration and notes that mathematical time units cannot be successive. Then, he offers a new definition of time units and creates his famous concept of duration. In this manner, he elaborates a philosophical concept of number. This concept is not created in *Matter and Memory* (1896). Bergson invented it much earlier (1881–3), but he continues to use it to describe the being of matter and perception.

In summary, the Bergsonian theory of perception[8] is not only psychological (theory of the nervous system) but, mainly, philosophical (theory of knowledge, theory of being). This theory provides an exemplary illustration of the specificity of philosophy. Indeed, philosophy could be defined as follows: (1) semantically analysing definitions of sciences (humanities and hard sciences) to identify obscurities, contradictions, etc. and (2) creating new definitions (philosophy is not a semantic critic of sciences) to resolve these difficulties.[9] Therefore, it could be concluded that philosophy is not an alternative to sciences, and sciences are not an alternative to philosophy. They are indispensable complements (Miquel 2007: 40).

Notes

1 Cognitivism is not computationalism. The cognitive system does not work as a computer. For example, the neural coding could be analogical and not digital (Nieder and Dehaene 2009). The information processing is frequently 'intuitive' and non-analytic (Brainerd and Reyna 2002), etc.

2 For instance, if elements of one object are too far apart, all humans perceive a multiplicity of objects rather than a single object (Gestalt law of proximity). If it is possible to perceive a difference between 10 grams and 11 grams, it will be possible to perceive a difference between 100 grams and 110 grams (psychophysical law of Weber-Fechner). For cognitive psychology, perception of one or many objects (Gestalt) and perception of one or many intensities (psychophysical) respect Gestalt and psychophysical laws. However, these laws are not sufficient to explain perception. For example, a hungry person could perceive only one element of the picture if this

element is food. In other words, perceived objects are not only seen but processed. Sensory information is not only created but selected, questioned, preferred, etc. before being perceived.

3   In a famous experiment, Skinner gives electric shocks to rats. The rats could interrupt the electric shocks by pressing a lever. Skinner notes that the rats learn to press the lever in order to interrupt the electric shocks. He concludes that a stimulus (electric shocks) involves a response (pressed lever). For cognitive psychology, the response of rats is cognitive. They learn that pressing the lever interrupts the electric shocks. In fact, electric shocks activate this memorized knowledge, this internal representation, and that is why the rats press the lever. In brief, for behaviourism, a stimulus triggers a response (stimulus ⇒response); for cognitive psychology, a stimulus triggers an internal representation, and the internal representation triggers a response (stimulus ⇒internal representation ⇒response).

4   In cognitive psychology, there are two types of memory: procedural and declarative. Procedural memory contains all motor skills (e.g. knowing how to bike). Declarative memory consists of semantic memory (e.g. 'Bergson is a philosopher', 'Neurons are cells') and episodic memory (e.g. 'Yesterday, I was in Tokyo. I listened to conferences on Bergson').

5   When Bergson writes that 'representations' are not in the brain, he wants to say that 'mental images' (conscious images) are not in the brain. The word 'representation' does not mean the same thing in Bergson's philosophy and in cognitive sciences.

6   Bergson always develops his ideas after examining scientific facts. He does not build his metaphysical theories first and then try to justify them by science.

7   For Bergson, there is an artificial spatial order but also a real order in the duration, 'a certain order in time' (*un certain ordre dans le temps*) (TFW: 99).

8   Some details of this theory are not discussed here: point P, material and perceptive planes of consciousness, etc.

9   For Bergson, the philosophical (and scientific) truth is always more or less probable, because this truth depends on a number of solved problems (François 2013: 304; Miravète 2014). Our definition of philosophy agrees with this probabilistic conception of truth.

References

Bergson, H. (1910), *Time and Free Will*, trans. F.L. Pogson, London: George Allen and Unwin.

Bergson, H. (1911), *Matter and Memory*, trans. N.M. Paul and W.S. Palmer. London: George Allen and Unwin.

Brainerd, C.J. and V.F. Reyna (2002), 'Fuzzy-trace Theory and False Memory', *Current Directions in Psychological Science*, 11 (5), 164–9.

Bruner, J.S. and A.L. Minturn (1955), 'Perceptual Identification and Perceptual Organization', *The Journal of General Psychology*, 53 (1): 21–8.

Bruner, J.S. and C.C. Goodman (1947), 'Value and Need as Organizing Factors in Perception', *Journal of Abnormal and Social Psychology*, 42: 33–44.

Fédi, L. (1999), *Le problème de la connaissance dans la philosophie de Charles Renouvier*, Paris: Harmattan.

Forest, D. (2012), 'Introduction à Matière et mémoire', in H. Bergson, *Matière et mémoire*, 5–46, Paris: Flammarion

Fraisse, P. (1949), 'L'influence des attitudes et de la personnalité sur la perception', *L'Année psychologique*, 51: 237–48.

François, A. (2013), 'Notes "La Philosophie de Claude Bernard"', in H. Bergson, *La Pensée et le mouvant*, Paris: Presses universitaires de France: 446–8.

Fujita, H. (2007), *La logique mineure dans l'œuvre de Bergson. Pour un vitalisme (non) organique* (unpublished doctoral thesis), Lille: University of Lille 3.

Leeper, R. (1935), 'A Study of a Neglected Portion of the Field of Learning: The Development of Sensory Organization', *Journal of Genetic Psychology*, 46: 41–75.

Levine, R., I. Chen and G. Murphy (1942), 'The Relation of the Intensity of a Need to the Amount of Perceptual Distortion: A preliminary Report', *Journal of Psychology: Interdisciplinary and Applied*, 13: 283–93.

Miquel, P.-A. (2007), *Bergson ou l'imagination métaphysique*, Paris: Kimée.

Miravète, S. (2012), 'La durée bergsonienne comme nombre spécial', *Annales bergsoniennes*, 5: 401–18.

Miravète, S. (2014), 'Notice de "La philosophie de Claude Bernard"', in H. Bergson, *La Pensée et le mouvant*, 389–91, Paris: Flammarion.

Miravète, S. (2017), 'Spencer, Renouvier: comment Bergson a-t-il inventé la durée?' in S. Abiko, H. Fujita and Y. Sugimura (eds), *Considérations inactuelles. Bergson et la philosophie française au XIXe siècle*, 219–32, Hildesheim: OLMS.

Montebello, P. (2003), *Nature et subjectivité*, Paris: Desclée de Brouwer.

Montebello, P. (2007), *Nature et subjectivité*, Grenoble: Jérôme Millon.

Nieder, A. and S. Dehaene (2009), 'Representation of Number in the Brain', *Annual Review of Neuroscience*, 32: 185–208.

Wright, E. (1992), 'The Original of E.G. Boring's "Young Girl/Mother-in-Law" Drawing and its Relation to the Pattern of a Joke', *Perception*, 21: 273–5.

9

What is the 'Thickness' of the Present? Bergson's Dual Perception System and the Ontology of Time

Yasushi Hirai
Keio University

1 Dual visual system hypothesis in contemporary cognitive science

Throughout the studies of agnosia, blindsight[1] and visual illusions, it came to be known that we seem to have two independent visual systems, and many research programs are ongoing to identify their physiological substratum. For example, the distinction of 'vision for action' and 'vision for perception' maintained by Goodale and Milner (1992) is meant to have its neural correlate on the bifurcation between the 'dorsal' and 'ventral' stream at the exit of the Primary Visual Cortex (V1) (see Figure 9.1). In the case of *blindsight*, another neural bifurcation is located at the superior colliculus that exists before the V1 (see Figure 9.2).

Whatever the physiological correlate may be, the most important thing, as a matter of fact, is that it becomes more and more probable that our bodies are capable of executing appropriate reactions without any assistance from conscious visual images. Blindsight patients can avoid balls or obstacles without having any visual representation. For example, D.F., who has lesions in the ventral stream, can no longer determine the orientation of a slit, though she can still smoothly put an envelope into it (Milner and Goodale 1995). A sound person, having an erroneous representation by the effect of the Ebbinghaus illusion (e.g. a circular object *appears* to him as bigger than another),

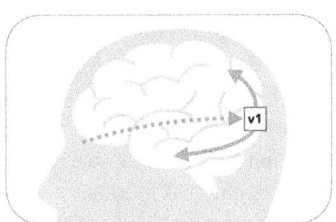

Figure 9.1 The diverging streams at the exit of the primary visual cortex. © Yasushi Hirai.

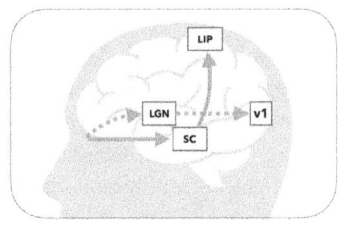

Figure 9.2 The diverging streams at the superior colliculus. © Yasushi Hirai.

would not be deceived in the movement of picking up these objects (i.e. the widths between fingers are correctly the same) (Haffenden and Goodale 1998). All this evidence strongly suggests the existence of a direct pathway from received stimuli to bodily reactions,[2] bypassing visual imaging processes.

To paraphrase this into the terminology of contemporary philosophy of mind, we can reasonably assume that some inferior animals are not equipped with 'phenomenal perception'.[3] Conversely, however, it turned out to be very unlikely that humans have nothing to do with 'functional perception'. Of course, we *can* and very often do act via images. Yet, this is not the only way of processing visual input. This fact leads us to discredit the opinion commonly shared among traditional theories of knowledge that one must not be able to execute any proper reaction to particular optical input without having any visual representation.

2 Bergson's dual perception theory

2.1 Automatic recognition as procedural perception

Consider the doctrine of dual perception developed in Chapter 2 of Bergson's *Matter and Memory*. It is well known that he distinguished two forms of memory: the memory of body which mechanically responds to situations (corresponding to today's 'procedural memory') and the image-memory which preserves particular events ('episodic memory'). Bergson then argued two forms of recognition based on these two memories. The first is what he calls 'automatic recognition', which operates during the execution of procedural memory. Because to 'recognize a common object is mainly to know how to use it' (MM: 101[111]), this recognition automatically executes processing of perceptual stimuli without the help of consciousness, or a representational image. Immediately after stimulations are inputted, corresponding actions are taken. This means that the perceptual information is safely relayed through to appropriate bodily response without any translation process into mental representation. Bergson stated that 'the body is capable [of this type of recognition] by itself, without the help of any explicit memory-image' (MM: 100[110]). For instance, one can walk around the accustomed tracks of one's hometown 'without having any distinct perception of the objects' (MM: 100[93]).

On the other hand, this 'distinct perception' (MM: 110[100], 124[112–13], 162–3[142]) or 'attentive recognition' refers to the other type of recognition based on pictorial representations. Although these two types of recognition can concur simultaneously in our experience, the former tended to be ignored by traditional theories of knowledge because only the latter is present for our conscious awareness. During that period, however, Bergson had already achieved their distinction through detailed analysis of disorders of memory and recognition. We can summarize the philosophical contributions of his theory of automatic recognition as follows:

1. Against the tradition of awareness-biased theory of knowledge, Bergson pointed out and verified the existence of non-representational, *procedural* perception on the basis of the study of pathological cases.

2. In addition, he showed that this type of perception is to be said more fundamental than the other from an evolutionary point of view, and conversely our image perception should be taken as an extra acquisition.
3. Bergson thus opened up the possibility to discuss image perception not as given *a priori*, but in its genealogy.

2.2 Attentive recognition as criticism of the reproductive model of perception

Let me make an additional remark on 'distinct perception', which corresponds to the mental representation. Its task is not to take a mental 'duplicate' of the real world, which would surely be a mystical process.[4] Our representational perceptions, far from being faithful copies of objective reality, consist in labours of supplement, processing and adaptation by our side (criticism of 'pure knowledge': MM: 16[24]).

According to Bergson's expression, our distinct perceptions are composed of 'two opposite currents, of which the one, centripetal, comes from the external object, and the other, centrifugal, has for its point of departure that which we term "pure memory"' (MM: 142[163]). What our bodily 'motor diagram' adopts from the external world is no more than a schematic outline of the displayed objects, whereas the rest of our perception consists of picking up the appropriate image materials from our memory (reduction), blending them into a proper image (contraction), and pasting it on the kinetically drawn scheme. The rich details that actually make up this pictorial world such as figures, textures, patterns, designs and so on are created through reusing and recycling our own past image materials. Our distinct perception yielded by this synthesis is far from an exact reproduction of the external objectivity. One literally sees, with eyes open, their own memories.

This can be further argued based on recent research in neurosciences. According to *Ganong's Review of Medical Physiology, 23e*, the number of the optic nerve is said to be 1,200,000.[5] This means that our visual organ ranks with a camera of 1.2 million pixels, which is desperately poor (recall that the first iPhone 2007 camera already had 2 million pixels). It is also reported that, among the numerous nerves coming into the V1, optic nerves from the retina occupy only 4 per cent, and 96 per cent of the fibres come from other various parts of the brain. In our phenomenal experience, exogenous ingredients take only a very small proportion.

To return to Bergson's attentive recognition, it is worth noting that the image-memories applied for rendering and decorating our perception are not used in its original form, i.e. as particular partial scenes of a particular past event. Firstly, several elements adaptable to the present requirement of perception (of the motor diagram) are selected from the vast stock of past images (1. selective reduction), then, secondly, regulate each other by means of interpenetration (2. fusional contraction) to be nicely integrated into actual sight.

This picture is an outcome of Bergson's theory of recollection. The point is that memories are not necessarily reproduced in the same alignment and timescale as the original; in other words, the same stored memories can be reproduced in different combinations, magnifications and resolutions at each recollection process.[6] In Chapter 3 of *Matter and Memory*, Bergson talks about the famous 'rotation and translation' (MM: 220[188]) movement through multiple 'planes' shaped as inverse cones in order to elucidate this point. It is also interesting that this dynamic rearranging ability of our memory may be linked to what is called the *gist* of today's cognitive sciences.[7]

Thus far, all that has been pointed out may seem to be only a retrospective coincidence. However, the truly inspiring part of this theory is the metaphysical background underlying all these psychological insights.

3 Direct perception theory: avoiding ontological detachment between things and perceptions

In Chapter 1 of *Matter and Memory*, Bergson argued the 'pure perception' as an abstracted ideal state devoid of any intervention of memory functions. Its aim is to ensure the direct realism showing that our perception reaches the objective reality, at least in its 'impersonal basis' (MM: 71[69]) upon which luxurious decorations are to be provided by image-memory. Let us begin by confirming how Bergson responds to the following two main reasons which motivated the ontological separation between things and perceptions: (1) 'epistemic relativity' (one and the same set of objects may produce different perceptions by individuals and by species), and (2) 'essential difference' (objects are physical and extensive, whereas perceptions are mental and inextensive).[8]

3.1 Epistemic relativity explained by the 'reduction system'

A *systems theoretic approach* has enabled Bergson to explain epistemic relativity while maintaining direct realism. One and the same element can be discriminated according to the 'system' to which it belongs. The material world and perceptual world coexist simultaneously in the same spatial region but are differentiated.

Systems are distinguished from each other according to their way of organization, or how to interrelate their constituents. A system has no centre (MM: 15[22]), and each component is 'related only to itself' (MM: 13[21]) or 'referred to itself' (MM: 12[20]). This is the system of 'matter' which belongs to Science. On the other hand, a 'perception' system is constituted in the following way:

[Constitution of reduction system of perception]

1. Take any one thing defined as a set of a finite number of possible reactions against the surroundings (call it 'my body'[9]) and adopt it as a 'centre'.
2. This set of limited possible reactions gives us a subset of the original system, which is now centred and reduced of irrelevant components.

Different 'my bodies' constitute different reduction-systems upon one and the same 'matter system'. There are, in principle, various way of centring, and that explains the variety of *Umwelt*, or diversity of perceptual space by species,[10] rather than exclusively by individuals.

> I call matter *the aggregate of images, and* perception of matter *these same images referred to the eventual action of one particular image, my body.*
>
> MM: 8[17]

> [T]he same images can enter at the same time into two distinct systems, one belonging to *science,* wherein each image, related only to itself, possesses an absolute value; and the other, the world of *consciousness,* wherein all the images depend on a central image, our body, the variations of which they follow.
>
> MM: 13–14[21]

In this picture, the distinction between objective universe and subjective worlds does not entail any ontological dissociation. Two systems may include *materially* the same common constituents, without losing their individualities as *systems*. Thinking of the General Systems Theory established by Ludwig von Bertalanffy (1901–72) more than fifty years later, it is not striking that this logical equipment itself might have prevented the proper understanding of the crucial insights of *Matter and Memory*. Here, we can formulate the following principle:

> [Principle of multiple constitution]
> '[T]he same images *can belong at the same time to two different* systems'.
>
> MM:13[20]

Accordingly, the material world and my perception are in whole-part relationship *in terms of their components*. However, each of them, *seen as a system,* is sufficient to itself and even independent from the others. Hence, another principle may be proposed as follows:

> [Principle of independence]
> '[N]either of the two systems of images is implied in the other, and each of them is sufficient to itself'.
>
> MM: 15[22]

It is crucial that partially common components can constitute multiple independent systems according to how they are organised. *Even if the one system* contains *all the elements of the other, the former does not* imply *the latter* (one cannot deduce the latter from the former). As Bergson stated, '[i]f you posit the system of images which has no centre ... I see no reason why to this system should accrue a second' (MM: 15[23]). Conversely, to fabricate the perceptual system from the material system, you must end up in 'conjur[ing] up some *deus ex machina*' (MM: 15[23]).

Based on this logical independency between systems, Howard Pattie, a famous contemporary systems theorist, introduced the concept of 'complementarity' as follows: 'I am using complementary here in Boltzmann's and Bohr's sense of logical irreducibility. That is, complementary models are *formally incompatible* but *both necessary*. One model cannot be derived from, or reduced to, the other'.[11] In the very same way, Bergson firmly denies the reducibility from one system to the other (dual criticism to the idealism and the realism: rejection of ontological detachment).

4. 'Extended Mind' not only in space, but also in time

As argued above, the 'reduction-system' can explain the epistemic relativity. However, the other reason for the essential separation, i.e. the (so-called) 'immaterial' nature of perception, especially of the sensible quality (*quale*), requires another explanation.

4.1 Essential difference explained by 'contraction'

It was the concept of a *temporally extended system* (and the *time-scale shift* brought about by it) which enabled Bergson to explain the appearance of essential difference without giving up the direct realism. In this picture, one can even *localize* in the material world sensible qualities usually considered to be inextensive. In this regard, Bergson states:

> For if we follow out to the end the principle according to which the subjectivity of our perception consists, above all, in the share taken by memory, we shall say that even the sensible qualities of matter would be known *in themselves*, from within and not from without, could we but disengage them from that particular rhythm of duration which characterizes our consciousness.
>
> MM: 75[72]

Each one of our (elementary[12]) sensible qualities has its spatial location. Further, the 'subjectivity' – or rather 'heterogeneity' according to Bergson's favoured expression – of our perceptual elements originates from the fact that our perception is extended not only in the dimension of space, but also of time. Justifiably, Bergson claims that the material system can be reduced to the 'continuously renewed present' as a single moment (the minimum amount) of duration. By contrast, each perception of ours contains, in an irreducible way, temporally extended elements. For example, to be able to distinguish two different colours, it is required to have a duration long enough for differentiating the wavelengths of inputted electromagnetic waves. Therefore, there should exist another principle which permits us to retain the previous moments that material systems cannot, and to include them in a single temporal unit for perception. This very principle is called *memory*.

Bergson names *memory* in general as the condition which enables antecedent moments to be carried over into posterior ones. It is important that it *may* indicate our mental ability of same name, but this definition in itself does not imply any mentality or personality.[13]

> [Introduction of '*memory*']
> We call *memory* the condition under which a system can be organized in a way extended in temporal dimension (with non-simultaneous components).
>> However brief we suppose any perception to be, it always occupies a certain duration, and involves consequently an effort of memory which prolongs one into another a plurality of moments.
>
> MM: 25[30–1]

> [M]atter, as grasped in concrete perception which always occupies a certain duration, is in great part the work of memory.
>
> MM: 237[202–3]

This shows that components of a perceptual system can hold a plurality of material moments thanks to *memory* of a higher degree of intensity than that of a material system, with the latter in turn being defined as the briefest possible succession (i.e. 'elementary memory'[14]).

How, then, can we produce the sensible qualities? The right response is that 'nothing *produces* them'. Recall the principle of multiple constitution: our sensible quality is not a final 'production' at the 'end' of a causal chain of physical process. As Bergson states, 'in the visual perception of an object the brain, nerves, retina *and the object itself* form a connected whole, a continuous process in which the image on the retina is only an episode' (MM: 285[241]). They are composed of the same constituents of the material system, but extended temporally, including its 'immediate past'. Therefore, all the ingredients necessary to constitute a sensible quality should already be given in this objective world, provided it is temporally extended. Nothing produces them, or in other words, they are produced only *through Time*.

That being said, *memory* entails another principle, which explains qualitative heterogeneity of these elements. How? By hypothesis, the 'matter system' as a whole consists exhaustedly in one single moment. That is to say that preceding moments can no longer exist *materially*. As for the ontology of matter, Bergson clearly takes the position of presentism. A material thing does not perdure, but endures. That is why we needed to introduce *memory*. Yet, how should we depict these new acquisitions of temporal extension?

> Depiction A (extended in an absolute time-based coordinate)
> The 'perception system' has a larger extent in the direction of the axis of time than that of the 'matter system'.

This type of picture should be supposed by Bergson when he talks about 'my concrete and complex perception - that which is enlarged by memories and offers always a certain breadth ['*épaisseur*' in French, literally means 'thickness'] of duration' (MM: 26[31]), as opposed to the 'mathematical moment' or 'ideal present'.

On the other hand, it was of highest importance for Bergson's thesis that *Time cannot be measured*, according to which we have no physical way of measuring the so-called absolute length of time (all we can do is to compare the relative intervals between instants): one can call this the 'immeasurability of time' thesis. Taking this into consideration, the depiction A might not be regarded as sufficiently adequate.

Thus, if we now decide to denote each of these intrinsic unit of time according to the term of 'moment',[15] we can translate the same fact as below: one single moment of our perception can contain more than one material moment. This is in stark contrast to the spatial inclusive relation of these two systems. Spatially, it was matter that contained perception.

By using the term of 'moment' defined in such a way, we are now able to convert the depiction A into the depiction B (see Figure 9.3):

Depiction B (expressing the difference as a *ratio*, not supposing the absolute time)
One perceptual moment contains more than one material moment.
Or, the former is of *higher density* than the latter.

Based on the depiction B, we finally obtain a series of central concepts such as contraction, tension, intensity, rhythm of duration and so on, each of which allows various degrees. In addition, since each of the multiple physical moments now constitutes altogether one single perceptual moment, it is necessary for the original moments to lose their 'unity' resolved into a new whole. This *fusion* triggers the emergence of the qualitative heterogeneity.

[Introduction of *contraction*]
We call *contraction* that which enables a system to contain its components in a relatively *dense* form.

[Principle of heterogenity]
As an effect of *contraction*, the original elements lose their discreteness, whereas the whole acquires a new emergent quality because of their temporal fusion.
> Between sensible qualities, as regarded in our representation of them, and these same qualities treated as calculable changes, there is therefore only a difference in rhythm of duration, a difference of internal tension.
>
> MM: 330[278]

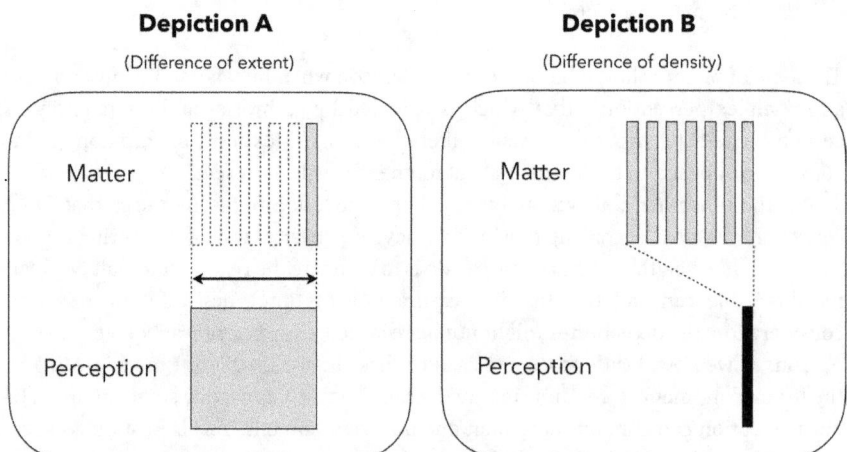

Figure 9.3 Depiction A and B of contraction. © Yasushi Hirai.

> The qualitative heterogeneity of our successive perceptions of the universe results from the fact that each, in itself, extends over a certain depth of duration, and that memory condenses in each an enormous multiplicity of vibrations which appear to us all at once, although they are successive.
>
> MM: 73[77–8]

In fact, one vibration of electromagnetic wave captured in a 'moment of matter' has no sensible heterogeneity.[16] However, *condensed* in a 'moment of perception', these same vibrations acquire a novel heterogeneous quality that they did not have in the coextensive material system. Thus, the sensible qualities consist only of objective elements,[17] but extended in the temporal direction, located in a certain place. Colour *qualia* can thus arise on the side of the material world. Needless to say, this will turn out to be a very singular statement in the contemporary philosophy of mind.

Let us make a brief remark on the difference between the 'retention' of the contemporary 'retentionalists'[18] and that of Bergson. His contraction theory is especially remarkable in the sense that it does not take any mental duplicate. It is *not* a psychological entity assumed as *representing* the immediate past but *the immediate past itself* that enters into the 'perception system'. If one adopts provisionally the depiction A, it shows us a clearly 'extensionalist' view of consciousness because each of our perceptual moments has its own 'thickness' of duration. On the contrary, under the depiction B, each perceptual moment is more densely compressed than a material one. This might mislead to an appearance of 'retenionalism'. Even in this case, however, the difference is to be of relative density, not that the duration of perception extends in an essentially different direction from that of material moments (in an orthogonal way (cf. Dainton 2013), as typically do retentionalists). Regardless, numerically same elements still constitute two distinct systems at the same time.

5. The existence of a virtual past

Bergson thus argues the temporal extension of our perception. The 'thickness' of our duration is contracted in a perceptual moment, which yields a new qualitative aspect. This is what is called 'my present' thesis, developed in Chapter 3 of *Matter and Memory* (MM: 176[152–6]). However, as it turns out, this 'my present', being stated to 'necessarily occup[y] a duration', still remains on the surface of a 'sensorimotor' mechanism (hence, we see that 'sensori-' may imply the timescale of subjective heterogeneity).[19] In that sense, we may understand why it is not strange that Bergson ascribes such modifiers as '*quasi*-instantaneous' (MM: 197[154]; 277[234]), '*so to speak* instantaneous'(MM: 79[68]) to this 'duration'. However, is this the end of the story? Of course not, because we have the other type of perception, i.e. the 'distinct perception' (attentive recognition), which originates with another function of *memory*: covering (or projection).

In short, memory in these two forms, *covering* as it does with a cloak of recollections a core of immediate perception, and also *contracting* a number of external moments into a single internal moment, constitutes the principal share of individual consciousness in perception, the subjective side of the knowledge of things.

MM: 25[31], emphasis added

Memory, inseparable in practice from perception, *imports* the past into the present, *contracts* into a single intuition many moments of duration, and thus by a twofold operation compels us, *de facto*, to perceive matter in ourselves, whereas we, *de jure*, perceive matter within matter.

MM: 80[76], emphasis added

5.1 Bergson's extended ontology and his theory of 'Hybrid' growing universe

According to this third dimension of perception, Bergson finally accepts a new type of entity, i.e. immaterial 'states', which are of the 'virtual past' and no longer of the 'immediate past'. His ontological commitment to the reality of 'states' (*états*) or 'events' (*événements*), distinguished from that of 'objects' (*objets*) or 'things' (*choses*),[20] is quite remarkable,[21] hence resulting in what I call the 'hybrid' growing universe theory of time, which consists of *present 'things'* and *past 'states of affairs'*. It is no longer a 'block', because he does not accept the existence of past 'things'.[22]

5.2 Direct Recollection Theory

Following the same line of reasoning as perception, we can quickly discuss Bergson's direct recollection theory, according to which what we recall are *not* (so-called) memories of representational nature stored in the present moment, but *ontological states of affair* in a remote past.[23] As we have already seen in section 2.2 of this chapter, there various adaptations intervene during the task of retrieving a particular component under the form of original event, which has once lost its individuality in a fusional contraction. Thus, in the very same way as that of direct perception, just because you admit direct recollection, it does not follow that you need to commit these fallacies that we should all recollect the same things, or that our recollection must be a faithful facsimile of the past event which you meant to recollect.[24]

Firstly, Bergson's famous 'automatic conservation of past' thesis provides us a whole collection of objective past of this universe.[25] It is true that every '*thing*' collapses. However, it is equally the case that nothing can alter, much less erase, any '*event*' once it has happened in a particular way. In that sense, the past survives by itself.[26] Secondly, by way of selective reduction, you will obtain a subset system which contains only 'your' past events. Thirdly, contraction and its heterogeneity principle give rise to your 'mind'.[27]

It is too easy to forget the principle of multiple constitution. It is not only the case that one and the same past event may enter into different minds, but further it is true that even in one single person, it takes various systems according to the way of their reconfiguration. For example, 'A word from a foreign language, uttered in my hearing, may make me think of that language in general or of a voice which once pronounced it in a certain way' (MM: 188[220–1]). It is this very 'voice', which enters together with all others, that relates to the language into the fusional image of 'that language in general'.[28]

> They take a more common form when memory shrinks most, more personal when it widens out, and they thus enter into an unlimited number of different 'systematizations'.
> MM: 188[220]

5.3 Intervention of past in attentive recognition

Let us now return to the distinct perception. It should now become much easier to grasp what happens according to this perception. In our *image-ful* perspective, all the decorations and ornaments with which we complete the poor 'sketch' supplied by motor diagram into a form of elegant high-resolution 'picture' are not simply made of our actual resources. Instead, it is literally the objective past itself that is 'imported' here.

One may feel puzzled by Bergson's seemingly misleading – even systematic – paraphrasing of 'past' and 'memory' in his book,[29] on suspicion of some possible category error. However, we may now comprehend why it is by no means a mere miswriting, because he did have, as we have seen, a theoretically justified reason for *numerically* (not qualitatively, of course) *identifying* a memory with its target, i.e. a past event: just as a perception and its object were ontologically the same element taken in two different systems.

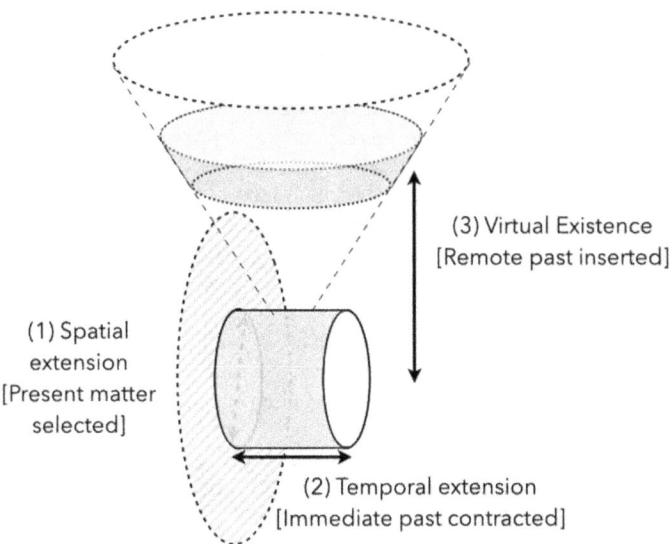

Figure 9.4 The triple structure of Bergsonian space-time. © Yasushi Hirai.

We have hitherto traced, step-by-step, the logical move of his original systems theoretic approach in *Matter and Memory*, and how crucially it works in defence of his direct perception and direct memory theory. This investigation revealed, we hope, a very singular, *threefold* spatio-temporal structure of our experience.

Our physico-phenomenal experience, not only extended in space, but furnished with tasteful qualities made of contracted immediate past moments, finally gains the richest power of imagery description through importing the vast virtual resources, i.e. remote past events preserved by the growing universe itself. Thus, our perception, differently from matter existing only in a present moment – and also from other animals possibly equipped with *qualia* but without any variably structured images – provides us a luxurious spatio-temporal construction *triply* articulated by spatial extension, temporal contraction, and virtual existence.

This conference is supported by the Grant-in-Aid for Scientific Research (B) under Grant No. 15H03154 from the JSPS (Japan Society for the Promotion of Science). Website: http://matterandmemory.jimdo.com/

Notes

1. In an international symposium held in Fukuoka in 2010, I discussed Bergson's recognition theory, adapting it to the context of blindsight. Cf. Hirai (2010), Weiskrantz (2009), and especially Yoshida and Isa (2015).
2. Limited to elementary, well-mastered and stereotyped movements.
3. E.g. Maeda (2008) refers to pigeons.
4. This involves a threefold criticism: (a) against 'pure knowledge', (b) against Epiphenomenalists, and (c) against centripetal interpretation of attentive recognition (Chapter 2 of *Matter and Memory*).
5. Barrett et al. (2011). Koch reports that the visual information stream conveys only 'a couple of megabytes per second' (Koch 2012: 55).
6. '[I]t is embraced in an intuition of the mind which I may lengthen or shorten at will; I assign to it any duration I please; there is nothing to prevent my grasping the whole of it instantaneously, as in one picture' (MM: 91[85]).
7. Koch (2012: 56): '*Gist* refers to a compact, high-level summary of a scene – a traffic jam on a freeway, crowds at a sports arena, a person with a gun, and so on. Gist perception does not require attentional processing: When a large photograph is briefly and unexpectedly flashed onto a screen while you're being told to focus on some itsy-bitsy detail at the centre of your focus, you still apprehend the essence of the photo. A glimpse lasting only one-twentieth of a second is all it takes. And in that brief time, attentional selection does not play much of a role'.
8. I have argued three types of dissociation including one that is 'causal' in my article (Hirai 2005).
9. This wording follows the expression of the philosopher at MM: 1[11].
10. A link to Jakob von Uexküll's theory. Cf. Hirai (2010), Miyake and Hirai (2018), Tani, Miyake and Hirai (2018).
11. Pattee (2000), as cited in Gazzaniga (2011: 122–3).

12 Those which were called 'representational' in TFW, Chapter 1. We shall criticize in the fifth section the mixing up of all types of *qualia* in the contemporary discourse of philosophy of mind. See note 29 below.
13 Bergson (2009: 46). Bergson firmly denies any 'anthropomorphic sense' (2009: 42) ascribable to this 'elementary memory' (2009: 42) or 'impersonal consciousness'.
14 See note 16.
15 However rapid we suppose it to be, pure perception, in fact, occupies a certain depth of duration, so that our successive perceptions are never the real *moments of things* as we have hitherto supposed, but are *moments of our consciousness* (MM: 75[71], emphasis added).
16 In reality, matter *does* have one at the lowest, i.e. insensible, degree. Cf. Chapter 4 of *Matter and Memory*.
17 In several contexts, Bergson counts them as 'share' (*apport*) of our subjectivity (MM: 25[31]). However, this subjectivity, as will be later revealed, in turn results in a form of virtual objective past.
18 Such individuals include Brentano, Hodgson, Kelly, James and Husserl, among others. See Dainton (2011, 2013, 2014a, 2014b and 2014c).
19 Having said that, the importance of this level of sensible heterogeneity (contraction on the sensory level) as the very beginning of our liberty (see Bergson 2013: 25 '*un commencement de liberté*') does not lose its appeal. 'If there are actions that are really free, or at least partly in determinate, they can only belong to beings able to fix, at long intervals, that becoming to which their own becoming clings, able to solidify it into distinct moments, and so to condense matter and, by assimilating it, to digest it into movements of reaction which will pass through the meshes of natural necessity' (MM: 279[236]).
20 There are other expressions such as '*souvenir*', '*expérience*', '*histoire*' and so on. However, there is no occurrence, to my knowledge, of using '*objets*' or '*choses*' for the remote past entity.
21 Further, he admits that they are 'immaterial' and 'inextensive' (MM: 180[156]).
22 This has the advantage of being able to avert the famous Braddon-Mitchell's attack (2004).
23 Fairly speaking, Bergson's expression still tends to have a more psychological tone in this book. For example, 'our complete perception is filled with images which belong to us personally, with exteriorized (that is to say recollected) images' (MM: 71[69]). However, our aim here is to try to reconstruct his system in the most consistent way possible.
24 This type of fallacy seems to risk confusing 'content' of recollection and its 'object', or numerical and qualitative identity.
25 On the topic of this logical connection between Event Ontology and Past, I believe that the work of Isashiki (2010) is of highest importance.
26 'This survival of the past *per se*' (MM: 193[166]).
27 By effect of condensation and its heterogeneity principle, this reduction-system (a subset of the total universal past) acquires an unpredictably novel quality, which is your 'mind', a contraction-system composed of virtual past elements, but in a fusional form.

'[O]ur character, *always present* in all our decisions, is indeed the actual *synthesis of all our past states. In this epitomized form (forme condensée)* our previous psychical life exists for us even more than the external world, of which we never perceive more than a very small part, whereas on the contrary we use *the whole of our lived experience*. It is true, that we possess merely a digest of it ...' (MM: 188[162], emphasis added).

'The whole of our past psychical life conditions our present state, without being its necessary determinant; whole, also, it reveals itself in our character, *although no one of its past states manifests itself explicitly* in character. Taken together, these two conditions assure to each one of the past psychological states a real, though an unconscious, existence' (MM:191[162], emphasis added).

'*Collecting, organizing the totality of its experience* in what we call its character, the mind causes it to converge upon actions in which we shall afterwards find, together with the past which is their matter, *the unforeseen form which is stamped upon them by personality*' (MM: 225–6[192], emphasis added).

Thinking that the central aim of this book was to *reformulate the traditional mind-body problem under the light of philosophy of time*, especially of the ontology of past, the point turns out still more conclusive.

28 Therefore, he seems to distinguish two types of *qualia* the *sensible, elementary qualities* such as colours, sounds, etc. on the one hand, and the *superior, sentimental ones* such as 'a confused sense of the striking quality or of resemblance' (MM: 171[206]), or 'the sense of familiarity' (MM: 111[101]). The latter does not derive from the immediate past.

29 For example, 'Memory ... imports the past into the present' (MM: 76[80]), 'the determination of the present by the past' (MM: 163[189]).

References

Barrett, K.E., S.M. Barman, S. Boitano and W.F. Ganong (2011), *Ganong's Review of Medical Physiology*, 23rd ed., trans. Y. Okada, Tokyo: Maruzen.

Bergson, H. (1911), *Matter and Memory*, trans. N.M. Paul and W.S. Palmer. London: George Allen and Unwin.

Bergson, H. (2007), *Matière et Mémoire*, Paris: Presses universitaires de France.

Bergson, H. (2009), *Durée et simultanéité : à propos de la théorie d'Einstein*. Paris: Presses universitaires de France.

Bergson, H. (2013), *Essai sur les données immédiates de la conscience*. Paris: Presses universitaires De France.

Braddon-Mitchell, D. (2004), 'How Do We Know it is Now Now?' *Analysis* 64: 199–203.

Dainton, B. (2011), 'Time, Passage, and Immediate Experience', in C. Callender (ed.), *The Oxford Handbook of Philosophy of Time*, 382–419, Oxford: Oxford University Press.

Dainton, B. (2013), 'The Perception of Time', in H. Dyke and A. Bardon (eds), *A Companion to the Philosophy of Time*, 389–409, Malden, MA: Wiley-Blackwell.

Dainton, B. (2014a), 'Flow, Repetition and Symmetry', in N. Oaklander (ed.), *Debates in the Philosophy of Time*, 175–212, London: Routledge.

Dainton, B. (2014b), 'The Phenomenal Continuum', in D. Lloyd and V. Arstila (eds), *Subjective Time: The Philosophy, Psychology, and Neuroscience of Temporality*, 101–37, Cambridge, MA: MIT Press.

Dainton, B. (2014c), 'Unity, Synchrony and Subjects', in D. Bennett and C. Hill (eds), *Sensory Integration and the Unity of Consciousness*, 32, Cambridge MA: MIT Press.

Dyke, H. and A. Bardon, eds (2013), *A Companion to the Philosophy of Time*, Malden, MA: Wiley-Blackwell.

Gazzaniga, M.S. (2011), *Who's in Charge?: Free Will and the Science of the Brain*, New York: HarperCollins.

Goodale, M. and D. Milner (1992), 'Separate Visual Pathways for Perception and Action', *Trends in Neurosciences*, 15: 20–5.

Haffenden, A. and M. Goodale (1998), 'The Effect of Pictorial Illusion on Prehension and Perception', *Journal of Cognitive Neuroscience* 10 (1): 122–36.

Hirai, Y. (2005), 'Images: A Radical Externalism of Perception' (in Japanese), *Tetsugakushi* 47: 39–54.

Hirai, Y. (2010), 'Interprétation Bergsonienne de la Vision Aveugle: Perception Motrice, Dissociation ou Indétermination?' (in French), International Symposium: Thought and Movement: Aristotle, Begson and Deleuze, held in Fukuoka, 27 March 2010.

Isashiki, T. (2010), *Metaphysics of Temporal Modality*, Tokyo: Keiso Shobo.

Koch, C. (2012), *Consciousness: Confessions of a Romantic Reductionist*, Cambridge, MA: MIT Press.

Maeda, T. (2008), 'Qualia of Robots, Zombies and Humans', in Shibata, Nagataki, Mino et al. (eds), *The Enigma of Consciousness and Qualia*. Showado.

Milner, D. and M. Goodale (1995), *The Visual Brain in Action*, Oxford Academic: Oxford.

Miyake, Y. and Y. Hirai (2018), 'Incorporating the Bergson Model into Artificial Intelligence', in Y. Hirai, H. Fujita and S. Abiko (eds), *Rebooting Bergson's Matter and Memory*, 120–38, Tokyo: Shoshi-shinsui.

Pattee, H.H. (2000), 'Causation, Control, and the Evolution of Complexity', in P.B. Andersen, P.V. Christiansen, C. Emmeche and M.O. Finnerman (eds), *Downward Causation: Minds, Bodies and Matter*, Copenhagen: Aathus University Press.

Tani, J., Y. Miyake and Y. Hirai (2018), 'Bergson and the Future of Artificial Intelligence', in Y. Hirai, H. Fujita and S. Abiko (eds), *Rebooting Bergson's Matter and Memory*, 139–74, Tokyo: Shoshi-shinsui.

Weiskrantz, L. (2009), *Blindsight: A Case Study Spanning 35 Years and New Developments*, Oxford: Oxford University Press.

Yoshida, M. and T. Isa (2015), 'Signal Detection Analysis of Blindsight in Monkeys', *Scientific Reports*, 5: 10755.

10

Affordance and Bergson

Tatsuya Higaki
University of Osaka

It is easy and convincing to see a certain connection between the theory of affordance and Bergson's thought. There seems to be an exact parallel between the fact that the theory of affordance was developed based on Gibson's visual psychology and became connected with evolutionary theory and discussions of the environment, and the fact that Bergson was strongly committed to the theory of visual images in his early thought and developed his biological ideas as an extension of that theory.

It may be interesting to place Deleuze in this context. Of course, Deleuze was strongly influenced by Bergson when he started his own thinking. In fact, even in the latter half of his life, he wrote *Cinema*, a theory of film, which could be interpreted as a commentary on *Matter and Memory* written in Deleuze's own way. Although there is a paradoxical twist in his work, in which he takes Bergson's criticism of cinema as leading to an illusion of a movement called 'cinematographic illusion' and transforms it into a thesis about cinema itself, the fact that Bergson's influence has culminated in the art of 'light' called cinema is proof of how one aspect of Bergson's argument is related to the visual (there are various discussions on the connection between affordance and Deleuze by Norihisa Kurenuma 2009).

I am not a good reader of Gibson, and I do not have more than a cursory knowledge of his theory of direct perception. I do know, however, that his theory of direct perception, in dealing with the problem of vision, forces us to make a fundamental change in our assumptions about vision. This leads to Bergson's claim about visual images, that what is seen is seen exactly in the place where it is seen. Needless to say, Bergson abolishes both subjectivism and objectivism in perception, and emphasizes that my interior and exterior are on the same footing as images (the preface to the seventh edition of *Matter and Memory* is very clear about this). It is true that the ideas about the texture of visual space and the detailed claims about the kinetic properties of moving and non-moving objects as discussed by Gibson cannot be found in Bergson. However, Bergson's thoughts on the subject and object of vision are very similar with those of Gibson. Also, the thesis of affordance that Kono Tetsuya uses as a slogan, that 'the mind is outside the body', may not correspond exactly to Bergson, but it is somewhat related to his expansion of the mental to cosmic scales in his monism of duration. In this sense, such a visual idea is certainly connected with the theme of holographic perception proposed by Stephen Robbins' chapter in this book.

However, when we set up the subject of Bergson and affordance, we have to keep in mind the gap between them. This is because affordance was formed from empirical science (not in a bad sense), while Bergson is a metaphysician. Therefore, even though the theory of affordance opens up a very precise view of perception and motion for those who support Bergson (for example, a detailed discussion of motion, perception, and environment such as that given by Masato Sasaki is not possible for Bergson, who depicts creative evolution), but its range seems to be limited to the soundness of empirical science. Bergson's theory of the image can only fix its gaze on a metaphysical plane (again, this is not a question of which is better or superior; it is only that the assumptions and directions of the approaches are different).

Let us consider the relation between Bergson and science. Bergson is often criticized (and has been criticized historically) for being too spiritualistic, but in fact he was a very scientific and positivist-minded. As can be seen in the fact that one of the sources of Deleuze's philosophy which he calls 'transcendental empiricism' is Bergson as well as Hume, Bergson is, by all accounts, an empirical theorist of science. This is evident in his psychology in *Essay*, his neuroscience and linguistics in *Matter and Memory*, his biology in *Creative Evolution*, and his sociology in *Two Sources*. Nevertheless, Bergson takes a sudden step away from empiricism. He moves towards a seemingly ridiculous argument with an almost innocent look on his face. To me, however close Bergson may be to the theory of affordance, he still seems unique in that he presents such an extraordinary metaphysics with such casualness.

It is clear that for Bergson, the discussion of perception is linked to that of memory through the brain. Yet Bergson's claims about memory are quite remarkable, for he says that in pure memory, the past remains as it is in itself. However, this is only a logical consequence of Bergson's view that time is a continuum that cannot be divided (or, more precisely, that dividing it changes its nature). If time is indivisible, then the past within the flow of time must be connected to the present in such a way that everything is not completely separated. Of course, the word 'memory' usually calls to mind our personal past. However, in memory, where everything exists as a continuous time, everything is continuous, so that not only all personal memories remain, but also all the images in this world must remain as memories. Where do they remain? Of course, it cannot be in the brain. The brain itself is an image, and perception is just an image within an image, but pure memory is the residue of all these images. Therefore, pure memory cannot be in the brain. Rather, the brain is a small portion of a virtually existing memory, a cross-section, and cannot possibly encompass it in the first place. This is also a precise indication that pure memory is outside the brain and all living organisms – the reverse side of the claim of the theory of affordance, that the mind is outside (pure memory I related to the brain only when it becomes an image of memory).

The entire past exists exactly as it was (virtually). Unscientific as it may seem, this is also a reasonable argument guided by the durational continuity of time (this is obviously so: it is only the evident claim that whatever existed exists in a certain way – virtually).

However, there is more to it than that. If duration is such a continuity, and if the past exists in itself, then the optical image that Bergson was so concerned with becomes something that *cannot be said to exist* in itself, just as a hologram is somewhat like an ephemeral image. In other words, the past is all there is, but the present, which is like a fragment of the flow of time, can be divided into any number of parts if it is taken alone, and cannot be said to exist (this is related to Zeno's paradox, which Bergson discusses in his early works, and is also indicated in his essay 'Memory of the Present and False Recollection,' which deals with déjà vu and is included in his *Mind-Energy*). Here, the present is regarded as something that does not exist, even though it had been emphasized so much as an image. Bergson, who had been talking about perception as an image, ends up saying that there is no present in which such an image can be found. This makes us wonder: what is going on?

If we only look at this, Bergson's philosophy certainly seems to contain some rather ridiculous metaphysical content. However, the problem is that these are precise consequences of his own thinking, and more importantly, Bergson gets these philosophical arguments from the data of science (the problem of déjà vu is taken from psychological data). So, if this argument is close to a metaphysical dream, it should be refuted by science, but science by itself does not have that power. And in my view, neither does the theory of affordance, which leads us back to habits even when discussing life forms.

Moreover, Bergson, who talks about evolution but ends up with superhuman beings, is much more scientific than your run-of-the-mill theories of evolution, isn't he? No matter how you look at it, human beings are not supposed to continue to exist beyond a certain point in time (which is natural from a scientific point of view). Can actual biology offer this vision? To what extent is it really possible to connect Bergson, who throws out such an extraordinary vision, with actual science? Shouldn't this point be properly reexamined?

Reference

Kurenuma, N. (2009), 'Mondai no Shingi to Jitsuzai no Kubun: Gibson to Bergson no Houhou', *Shiso* (1028): 171–92.

Part Three

Time and Duration

11

Neutral Monism, Temporal Experience and Time: Analytic Perspectives on Bergson

Barry Dainton
University of Liverpool

To say that Bergson's work has been largely overlooked in recent analytical philosophy would be an understatement. I will not hazard a guess as to whether this will change in the years to come, but one thing is clear: this situation *ought* to change. Some of the issues close to the centre of current analytical concerns – issues concerning the relationship between the mental and the physical, the nature of time, the ways in which consciousness is temporal – Bergson not only had interesting things to say, some of what he had to say may well turn out to be true. However, the traffic is not one-way. If analytic philosophy has much to learn from Bergson, I think it is also true that Bergsonians have something to learn from analytic philosophy. Recent discussions concerning the character of temporal experience shed useful light on at least one central aspect of Bergson's system. Bergson claimed that the nature of *durée* could be intuited, but not analysed. I will be arguing that *durée* is more susceptible to analysis that Bergson believed possible. Then there is the question of the nature of time itself. Attempting to situate Bergson's views within the framework elaborated in recent years by analytic philosophers is an intriguing exercise: it is far from obvious *where* Bergson can be situated in this framework.

Early (and recent) analytic philosophy: Bergson's unacknowledged role

I will start with some reflections on Bergson's main topic in *Matter and Memory*: the relationship between consciousness and the physical world.

According to Michael Dummett, the distinguishing feature of analytical philosophy from other schools is the belief 'first, that a philosophical account of thought can be attained through a philosophical account of language, and secondly, that a comprehensive account can only be so attained' (Dummett 1994: 4). Although the idea that analytic philosophy is primarily concerned with the analysis of language is widespread in *non*-analytic circles, it is by no means the whole story. In his 'How Did

We Get Here From There? The Transformation of Analytic Philosophy', Timothy Williamson – the current occupant of Dummett's Oxford chair – observes that from the 1970s onwards, Dummett found himself 'fighting a more global trend in analytic philosophy: a move away from the philosophy of language to the philosophy of mind ... many analytic philosophers were starting, heretically, to treat the philosophy of mind as more fundamental than the philosophy of language' (Williamson 2014). And this is true. What is also true – although most textbooks on the subject tend to overlook the fact – is that over the course of its brief history, philosophers in the analytic tradition have subscribed to a diverse range of sometimes quite exotic views on the nature of mind, and its place in the physical world.

According to the standard textbook story, analytic philosophy of mind was dominated until recently by physicalist and reductionist views. The story begins in the 1930s and 1940s, when behaviouristic approaches to the mind – inspired in part by Quine and Wittgenstein – were in favour. Over the next half-century these were abandoned for more promising and plausible alternatives: first, the mind-brain identity theory, then the computer-inspired functionalism of Putnam. Drawing on the cognitive neuro-sciences, analytic philosophers of mind in the 1980s and 1990s were perfectly happy to appeal to 'internal' states, processes and mechanisms. In their final chapters the more up-to-date textbooks acknowledge that by the mid-1990s, in some quarters at least, disenchantment was starting to set in. A number of philosophers – e.g. Nagel, McGinn, Strawson, Searle, Chalmers – expressed doubts regarding the dominant physicalistic reductionist accounts of phenomenal consciousness. By the 2000s, the doubts had increased significantly, and the hunt was on for non-reductive accounts of the relationship between consciousness and the physical world. One prominent (early) contender was the 'naturalistic dualistic' outlined by Chalmers in *The Conscious Mind* (1996). For Chalmers, conscious states and properties are non-physical particulars that are nomologically correlated with information processing in the brain and other physical systems.

Although it has its advantages, naturalistic dualism is not to everyone's taste: it bifurcates reality into two fundamentally kinds of ingredient, it also renders our consciousness epiphenomenal (or close to it). What are the alternatives?

What makes the problem of consciousness so very difficult is the conception of the physical world that the Scientific Revolution bequeathed to us. The only properties the elementary ingredients of the physical world – the electrons, quarks, gluons etc. – possess are properties such as mass, size, momentum, spin and charge: such entities most definitely do *not* possess phenomenal properties such as colour or sound (as experienced). Given this, it's difficult to see how any physical thing – our brains and neural processes included – *could* possess phenomenal properties. How could intrinsic qualities such as colour and sound (as experienced) emerge from collections of elementary particles that are entirely lacking in such qualities? There is a route out of this seemingly hopeless impasse. Perhaps electrons and quarks possess properties over and above those ascribed to them by current physics: perhaps they themselves possess phenomenal qualities by virtue of being themselves conscious beings (of a primitive kind). The desire to avoid dualism has thus led Galen Strawson and others contemporary philosophers to embrace panpsychism, the view that *everything* in the physical world is conscious to some degree.[1]

Reconceptualizing the nature of matter – with a view to replacing the austere primary-quality-only account bequeathed to us by the Scientific Revolution – as a means of overcoming dualism is by no means a novel approach to the problem of consciousness. Indeed, in contemporary analytic circles the strategy of inserting the phenomenal into the micro-physical has become known as 'Russellian Monism', since Bertrand Russell advocated the same move in the 1920s.[2] However, there is another way of bringing about the required *rapprochement*, one that was also – at times – advocated by Russell.

If you believe the immediate objects of ordinary perceptual experience are merely representations in the brain – in the manner of Descartes and Locke – then that's where phenomenal properties which feature in our ordinary perceptual experience will reside. If, in contrast, you think we are directly aware of external objects themselves (in the way we seem to be), then *that* is where the qualities in question will reside. When I look at a green apple, what I am perceiving is the apple itself (the physical object) and the green I'm seeing is a property of the apple, a property it has all the time, not just when someone is looking at it. This view of consciousness and perception was advocated by the American 'New Realists' in the early twentieth century – Holt, Marvin, Perry, Pitkin, Spaulding, Montague – and they favoured it precisely because it narrowed the gap between consciousness and the world. According to Holt,

> We are little tempted to believe that the colour of a flower fills all the space between the flower and the eye: and neither less nor more does it fill, or enter into, the peripheral nerves and skull. The entity responded to is the colour out there, two factors which involve two factors of response, but *that colour out there* is the thing in consciousness selected for such inclusion by the nervous system's specific response. Consciousness is, then, out there wherever the things specifically responded to are.... Mind and matter consist of the same stuff, and the little entity that in aggregates of various densities constitutes the secondary qualities is not far removed from the little atom that constitutes physical bodies ... their being included or not included in the class of things which we name a consciousness, depends for them both alike, in their being specifically responded to by a nervous system. But consciousness is in no sense at all *within* the nervous system.
>
> Holt 1912: 354

For Holt, a state of consciousness at a given time is a cross-section of events in the outer world (and the body) that are unified by virtue of being selected by a nervous system.

The New Realists did not emerge from a vacuum. During this period, the doctrine which would become known as *neutral monism* – characterized by Russell as the view 'that the raw material of the world is built up not of two sorts, one matter and the other mind, but that it is arranged in different patterns by its inter-relations, and that some arrangements can be called mental, while others may be called physical' (Russell 1921: 10) – was very much in the air. The New Realists took their inspiration from William James, who had started advocating the view in his 'Does Consciousness Exist?' (1904).

Of course, before James there came Bergson, whose *Matter and Memory* had appeared in 1896, nearly a full decade earlier. In his 1910 introduction to the latter,

Bergson tells us that the key to a better understanding of the relationship between the mental and the physical is appreciating that we need a new (to philosophers) way of conceiving of the physical. Matter, on the view he will be recommending, is an aggregate of 'images' and these images are different from both the mind-dependent 'ideas' of Locke and the quality-free 'matter' of Descartes. As Bergson goes on to make clear, one consequence of conceiving of matter in this way is that the objects we perceive really do have the properties (such as colour and temperature) that they seem to have – he embraces the same kind of perceptual realism as the New Realists. He did not draw back from the radical consequences of this realism: 'even physically, man is far from merely occupying the tiny space allotted to him ... For if our body is matter for our consciousness, it is co-extensive with our consciousness, it comprises everything we perceive, it reaches as far as the stars.... we are really present in everything we can perceive' (TS: 221–2).

In part IV of *Matter and Memory* we learn more of Bergson's cosmology. At bottom, the material world consists of cosmos-spanning fields consisting of 'numberless vibrations all linked together in uninterrupted continuity, all bound up with each other and travelling in every direction like shivers running through an immense body' (MM: 299). We know that for Bergson, duration (or *durée*) is the essential and distinctive feature of consciousness. It turns out that nothing in this vast physical field – not even its tiniest or briefest constituent – is entirely devoid of this vital property: there are only differences in rates of vibration and internal tension. An essential similarity between macro- and micro-scale properties is thus established: both essentially involve *durée*.[3]

Bergson's metaphysical vision – as Barnard (2011) aptly characterizes it – is a remarkable one, but it is also underdeveloped; Bergson himself does not pause to fill in much by way of detail, and so a good deal remains unclear. How are the simple forms of consciousness enjoyed by elementary physical particles (or fields) related to far more complex forms of human consciousness? If we see objects just as they are, how can the same object appear differently to different subjects? To what extent is Bergson's direct realism undermined (or enriched) by his claims relating to the extent to which memory influences the character of our perceptual experience?[4] While it would be good to have answers to these (and other) questions, however, it is not clear that Bergson's system is any *less* fully-worked out than those developed by the other neutral monists working a century ago.

What certainly *is* clear, however, is that Bergson has not been given his proper due by analytical philosophers as one of the originators of neutral monism. By way of an illustration: Stubenberg's Stanford Encyclopedia entry on the topic runs to over 30,000 words, and Bergson's name does not occur once.[5] In contrast, William James's name occurs frequently, as does Russell's. Although Russell had a dim view of Bergson's contribution to philosophy, it is ironic that Russell himself ended up embracing a position on the mind-body relationship very similar to Bergson's – albeit twenty-five years later! In his *The Analysis of Mind* we find Russell saying this:

> The stuff of which the world of our experience is composed is, in my belief, neither mind nor matter but something more primitive than either. Both mind and matter

seem to be composite, and the stuff of which they are compounded lies in a sense between the two, in a sense above them both, like a common ancestor.

Russell 1921: 10–11

Just a few years later, in his *The Analysis of Matter*, we find Russell saying this:

> We shall seek to construct a metaphysics of matter which shall make the gulf between physics and perception as small... as possible. We do not want the percept to appear mysteriously at the end of a causal chain composed of events of a totally different nature; if we can construct a theory of the physical world which makes its events continuous with perception, we have improved the metaphysical status of physics, even if we cannot prove more than that our theory is possible.
>
> Russell 1927: 275

The similarities between this and the position advocated by Bergson in 1896 are obvious. One of Bergson's main aims in *Matter and Memory* was precisely to narrow the divide between the experiential and the physical: both consist, ultimately, of different rhythms and densities of *durée*. Or as he puts it in the book's concluding section:

> Only one hypothesis, then, remains possible; namely that concrete movement [i.e. of material things] capable, like consciousness, of prolonging its past into its present, capable, by repeating itself, of engendering sensible qualities, already possesses something akin to consciousness, something akin to sensation.
>
> MM: 329

Since Russell was acquainted with *Matter and Memory*, it is difficult to believe that he was unaware that in advocating this sort of solution to the problem of consciousness he was following a path already taken by Bergson.[6] In contrast, it is quite easy be believe that Russell was very much aware of the similarities between his newly adopted stance and Bergson's, but preferred not to publicize the fact. Russell was, of course, hostile to the 'anti-intellectualism' he attributed to Bergson. Russell's rejection (from the 1890s) of Bradley's idealism meant that he was equally hostile to holistic metaphysical systems which stressed the interconnectedness of all things.[7]

As for James, he was more than happy to acknowledge the originality and significance of Bergson's contribution. In a letter to Bergson he wrote:

> [*Matter and Memory*] is a work of exquisite genius. It makes a sort of Copernican revolution as much as Berkeley's Principles of Kant's Critique did, and will probably, as it gets better and better known, open a new era of philosophical discussion. It fills *my* mind with all sorts of new questions and hypotheses, and brings the old into a most agreeable liquefaction.... I believe the 'transcendency' of the object will not recover from your treatment.
>
> James 1920: 180

I think something else is clear as well: Bergson's solution to the problem of consciousness is not solely of historical interest. For anyone interested in a unified, monistic, worldview, one where the mental and the physical are not fundamentally different in kind, then the approach pioneered by Bergson – a novel combination of direct realism and panpsychism – is well worth considering. It may be problematic in certain respects, but so too are all the alternatives.

By way of an illustration, in a recent survey of current work on the problem of consciousness Chalmers suggests (a) that Russellian Monism is one of the most promising solutions, and (b) James's 'combination problem' remains the most serious objection to it. If the panpsychist is to avoid a miraculous-seeming emergence of the phenomenal from the physical, then our rich and complex and complex conscious states must be wholly constituted from the more primitive states of consciousness belonging to the elementary particles which make up our brains. However, do we have any idea at all of how many conscious minds can *combine* or *agglomerate* to form a more complex consciousness? In his *Principles of Psychology*, James influentially suggested that we don't: 'Take a hundred of [the elemental minds], shuffle them and pack them as close together as you can (whatever that may mean); still each remains the same feeling it always was, shut in its own skin, windowless, ignorant of what the other feelings are and mean' (James 1890: 160). In view of the seriousness of this problem Chalmers suggests that a position he calls *panqualityism* may well be a more promising option:

> Instead, we would intuitively say that I am aware of redness, and that phenomenal properties involve awareness of qualitative properties. Likewise, phenomenal properties are always instantiated by conscious subjects, but qualities need not be. We can certainly make sense of the idea of a red object that is not a subject of experience.
>
> Chalmers 2016: 41

The key idea here is that if qualities such as colour and sound can exist independently of conscious *subjects*, then the problems of how such qualities combine to form complex states of the sort we are acquainted with can be addressed without supposing that subjects themselves, with their irreducibly distinct subjective points of view, can combine or fuse.[8]

Evidently, anyone who endorses panqualityism is not a million miles from Bergson, who in *Matter and Memory* argues that allowing sensory qualities to exist in the external world is essential if dualism is to be avoided. However, it should be noted that the problem of how the qualities combine or unite to form complex conscious states is by no means entirely solved by removing subjects from the picture. If the qualities reside in solely elementary particles, and the latter are spatially separated – as they invariably will be in ordinary matter, as ordinarily conceived – then it is by no means obvious that organizing these particles in brain-like configurations would suffice to unify the qualities in the required way. Why should it? There is a further story to be told here.

In the visual realm at least, this problem largely vanishes if we follow Bergson and combine panpsychism with a version of direct realism. For Bergson, objects have the properties they seem to have.⁹ The redness of a rose exists on the surface of the rose – it is not confined to mental representations generated in the brain (or immaterial soul). Since the redness on the surface of the rose seems to be a continuous unified expanse of colour, that is how it is in reality. In this sort of case at least there is no unifying of micro-experiential constituents to be done; the unity is out there in the world.¹⁰

Durée (partially) demystified

In a letter to Harald Høffding, Bergson observed that any account of his work which does not 'continually return to ... the very central point of the doctrine – the intuition of duration ... the point whence I set out and to which I constantly return is, on some level, a distortion'.¹¹ Given the central role *durée* plays in Bergson's thought we obviously need to understand what he's referring to if to understand his thought. Needless to say, Bergson has a good deal to say on this topic, here is Barnard attempting to summarize:

> *durée* is Bergson's term for the dynamic, ever-changing nature of consciousness, a consciousness expressed and manifested in-and-through-as *time*. From Bergson's perspective, *durée* is an indivisible fusion of manyness and oneness; it is the ongoing, dynamic, temporal flux of awareness; it is a flowing that is ever new and always unpredictable; it is the continual, seamless, interconnected, immeasurable movement of our awareness, manifesting simultaneously, as both knower and what is known.
>
> Barnard 2011: 6–7

There's a great deal to take on board here. To make things more manageable I'm going to focus on just some aspects of what Bergson took *durée* to be and ignore several others.¹² Whatever else it is, *durée* involves the *experience of change*, and it's this I want to consider.

Bergson is right: change does seem to be part of our ordinary everyday experience. If I turn my head or move my eyes, I see the surrounding world sliding smoothly by; even if I am sitting immobile in a chair I am aware of the unfolding of my inner soliloquy, and the continuing background flow of my bodily sensations; if I am listening to some music, I hear each note flow into the next – I hear the succession of notes *as* a succession – and each note itself has some finite duration, a duration I hear it persisting through. However, if change seems to be a ubiquitous feature of our experience it can also seem puzzling how it can be. The problem stems from three claims, each of which seems very plausible.

1. Any sort of change takes some amount of time to occur.
2. Our immediate consciousness is confined to the present. We can remember the past and anticipate the future, but we can only experience (directly) what is happening *now*.
3. The *now* has no duration. The present is what divides and distinguishes the past and the future; if we suppose the present has duration, then some parts of it will be earlier than others, and so not present after all.

Hence the problem: if the present has no duration, and our consciousness is confined to the present, how is it possible for us to experience change, given that change takes time?

Psychologists in the nineteenth century devised a way of circumventing this apparent paradox: the solution was to distinguish between the genuine or strict present which *is* durationless, from the 'psychological' or 'specious present' – to use the familiar terminology adopted by James in the *Principles of Psychology*. Since the specious present is *not* durationless (or seems so), it is perfectly capable of housing the change we encounter in our experience. As far as I am aware Bergson himself did not use the term 'specious present', but he evidently followed the same path:

> No doubt there is an ideal present – a pure conception, the indivisible limit which separates past from future. But the real, concrete, live present – that of which I speak when I speak of my present perception – that present necessarily occupies a duration.
>
> MM: 176

We have made some progress, but there is more to be done. For the specious (or 'real, concrete, live') present to constitute a genuine solution to the puzzle of experienced change, we have to be able to provide a metaphysically coherent account of it. Interestingly, in the literature on this topic, one can find a number of accounts, some of which differ fundamentally. Intriguingly, it is not immediately evident which of these accounts (if any) Bergson would have leant towards.

At the start of my 'Temporal Consciousness' (Dainton 2010a) entry for the *Stanford Encyclopedia of Philosophy*, I outlined three competing views of the nature of temporal experience. According to the first, which I labelled 'the Cinematic model' our streams of consciousness are composed of dense successions of momentary (or near-to-momentary) states, the contents of which are themselves momentary and entirely static. Our experience of change is the result of these static momentary states following on from one another in rapid succession. In Figure 11.1 below only a few of these momentary states are depicted – by the vertical lines – in reality there would be many more.

Two things are, I think, very clear: Bergson rejected this model (indeed, he almost invented the term for it), and he was right to do so. Since each momentary episode of

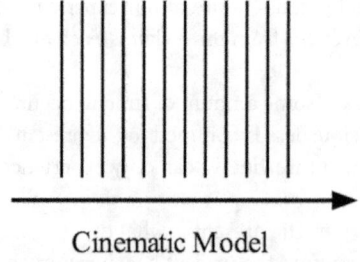

Figure 11.1 Cinematic model. © Barry Dainton.

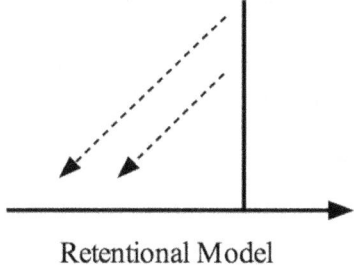

Figure 11.2 Retentional model. © Barry Dainton.

consciousness is entirely discrete and unconnected to its neighbours, there would be no *experienced* continuity here at all, and this is phenomenologically inaccurate: as Bergson rightly stressed, our consciousness is deeply continuous, each phase is felt or experienced as flowing into the next. No such flow is possible on the Cinematic model. Moreover, since the contents of each momentary state are themselves entirely static and self-contained, this model renders it impossible to directly experience change: it allows for a succession of experiences, but no experience *of* succession. Again, as Bergson rightly emphasized, it is impossible to generate experiences of dynamic motion – the kind we find in our experience all the time – from collections of immobilities.

The Retentional model, depicted in Figure 11.2, is in one respect similar to the Cinematic model, but in other respects very different. We are again supposing that a stream of consciousness is composed of a dense succession of momentary (or near-momentary) states – represented in the diagram by a solitary vertical line. However, the contents of these momentary states are *dynamic* rather than static: these contents seemingly extend through a brief interval of time, and they present (or represent) experienced change and succession.

As James's discussion in chapter 15 of the *Principles* reveals, this sort of account was endorsed by many German psychologists in the nineteenth century. Brentano was one of the first philosophers to advocate a variant of it, and when Husserl started thinking about time-consciousness he worked within the same framework. According to Husserl, a momentary time-slice through our streams of consciousness reveals a momentary 'primal impression' and an accompanying system of 'retentions' of our just-past experiences. In the diagram these retentions constitute the vast bulk of the vertical line, and the dotted arrows serve to indicate the regions of the recent past that they represent (the point at the bottom of the vertical line represents the momentary primal impression). Quite how retentions should be conceived is a matter of controversy among Retentional theorists – Husserl's views evolved over the course of his career – but they are generally assumed to be a distinctively vivid form of involuntary memory.

Why assume that our consciousness of change is confined to a momentary present? One consideration – which influenced Brentano – is the reality of those regions of time other than the present. Anyone who is convinced that reality itself is confined to a momentary present – and hence that the past and the future are wholly unreal – will have little option but to embrace Retentionalism. However, there is a further and different motivation. Thanks in part to Kant's influence, philosophers and psychologists in the nineteenth century were keenly aware of the distinction between a succession of experiences and an experience *of* succession. Let's suppose our succession takes the form of A-B-C, which we can take to be the adjacent phases of a perceived movement (or a continuously experienced sound). For an experience of succession to occur, it isn't enough for the contents representing A, B and C merely to occur in a succession. For them to be experienced *as* a succession the contents must be apprehended *together*, in a single unified awareness. In the absence of this, there would be no awareness of the succession, and hence no experience of it either. It was generally agreed (during this period, at any rate), that only contents that occur *simultaneously* can be presented together to a unified awareness. If simultaneity is required for experiential unity, the Retentional model – in some form or other – is inevitable.

What I have called the *Extensional model* is very different. In this view, our episodes of experiencing are themselves temporally extended, and so able to incorporate succession and persistence in a natural and direct way. For the Extensionalist, a specious present – represented by the blue rectangle in Figure 11.3 – is a temporally extended 'block' of experience, one whose parts are all experienced together as unified, but also as occurring in succession. If the content of this specious present is A-B-C, then just as in the Retentional case, we experience A flowing into B, and B flowing into C. (Or to put it a different way, A, B and C the neighbouring tones are experienced as 'diachronically co-conscious'.) The difference is that this experiential complex is itself extended through time – its components do *not* exist in a momentary conscious state.

William Stern was (perhaps) the first to explicitly advocate the Extensional model, in his 'Mental Presence-Time' in 1897.[13] Stern could see no reason why phenomenal unity had to be confined to contents that were (objectively) simultaneous: 'The psychic occurrence that takes place within a certain stretch of time may possibly form a unitary coherent act of consciousness, regardless of the non-simultaneity of its constituent parts.' Stern also finds fault with Retentional approaches. What we are looking for is a way of making sense of our *perception* of change. Retentionalists locate the perception of change in memory-like *reproductions* of previously experienced contents? By allowing properly perceptual contents to be experienced together, the Extensional model provides for precisely this.

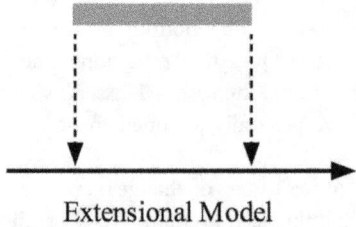

Extensional Model

Figure 11.3 Extensional model. © Barry Dainton.

There is, of course, more to be said about both approaches – see Dainton (2010a) for further elaboration – but we already have enough in view to pose the question I'm interested in. Of these two approaches to temporal experience, which would Bergson have preferred? Since Retentionalists and Extensionalists both purport to accommodate directly experienced change and succession, from a phenomenological perspective it might seem that there is nothing to choose between them, and hence no reason for Bergson to take a stand on the issue. However, on closer inspection grounds – powerful grounds – for a preference emerge.

Throughout his writings on these topics Bergson stressed the *continuity* of our experience. Every part of our experience flows, and it does so without break or interruption. Writing to James, Bergson observes 'when I wrote my essay on *Les Données de La Conscience* ... I was led, through an analysis of the idea of time, to a certain conception of psychological life which is entirely compatible with yours (except perhaps that I see *places of flight* in the *resting places* themselves ...' (Ansell Pearson and Mullarkey 2002: 357). In *L'Energie Spirituelle* he tells us 'I believe that our whole psychological existence is just like this single sentence, continued since the first awakening of consciousness, *interspersed with commas, but never broken by full stops*' (ME: 70).

Confining as it does our consciousness to discrete momentary experiential episodes, the Retentional model only partially accommodates the continuity of consciousness. Within a single specious present there is no difficulty: the contents of these brief intervals unfold in a fully continuous manner, just as they seem to. However, contents occurring in *different* specious presents cannot be experienced together, no matter how close together in time they might be. Suppose, for example, that the four syllables of 'articulate' are divided between two entirely separate specious presents thus:

[art-ic] [ul-ate]

The first syllable flows into the second, and the third flows into the fourth (as indicated by the hyphens), but there is no experienced flow linking the second and third syllables – these two experiences exist in entirely separate and disconnected conscious episodes. These ruptures in experiential continuity are a straightforward consequence of the Retentional models architecture, and it is not easy to see how this defect can be remedied.[14]

The situation with the Extensional model is very different, or at least, it is if this is developed in one particular direction. Consider these two Extensional specious presents – let's call them SP1 and SP2 – each of which – for convenience – has been divided into five successive phases, which we'll label 'P1', P2' and so forth. SP1 is a unified ensemble, and so P1-P2-P3-P4-P5 is an experienced succession, and similarly for the phases P6-P7-P8-P9-P10, which jointly constitute SP2. However, since SP1 and SP2 are entirely separate and non-overlapping experiences, there is no direct experiential connection between them. As a consequence, there is a rupture in the continuity of this stream of consciousness: there is no experienced continuity or flow connecting phases P5 and P6. Indeed, given the total rupture in experiential continuity, it is arguably wrong to view SP1 and SP2 as jointly constituting a *stream* of consciousness at all.

```
SP1              SP2
[P1-P2-P3-P4-P5] [P6-P7-P8-P9-P10]
```

There is a simple remedy available to the Extensional theorist. To restore the missing continuity we need simply hold that neighbouring specious presents overlap by sharing parts. This is illustrated below, where SP1*-SP5* are five specious presents which are related in this way.

```
SP1*    [P1-P2-P3-P4-P5]
SP2*     [P2-P3-P4-P5-P6]
SP3*      [P3-P4-P5-P6-P7]
SP4*       [P4-P5-P6-P7-P8]
SP5*        [P5-P6-P7-P8-P9]
```

As can be seen straight away, in this scenario P5 *is* experienced as flowing into P6: this happens in SP2*, when P6 first enters the scene. Since every phase (except the first and last) is experienced as flowing into the next, we now have as much continuity as is required for phenomenological adequacy.

There may, however, seem to be a problem: aren't contents being experienced many times over? Stream-phase P2, for example, occurs in SP1* and SP2*. If P2 *were* experienced twice over, this would be problematic from a phenomenological point of view. In fact, due to overlap via part-sharing, there are no such repetitions. The P2 stream-phase in SP1* is numerically identical with its counterpart in SP2*. There is just *one* (very brief) experiential episode, an episode that exists in a number of different larger wholes. P2 is no more experienced repeatedly than your experiences on New Year's Day were experienced many times over by virtue of being included in several different week-long periods of time (e.g. 26 December–3 January, 27 December–4 January, 28 December–5 January). In like fashion, stream-phase P5 occurs in S1*, S2*, S3*, S4* and S5*, but since we are dealing with the same token experience in each case, the P5 content (whatever it may be) is only experienced a single time, and not repeatedly. When P5 is experienced in SP1* it is the most recent addition to the succession P1-P2-P3-P4. It then remains in consciousness as P1 drops out of consciousness, and P6 (then P7, then P8, then P9) successively enter consciousness – until it too drops out of consciousness.

There is a further point of clarification to note. Let's suppose that SP1* contains the experiencing of a brief stretch of motion – e.g. its content is similar to what you would experience visually if you watched a football fly into the back of the net after being kicked by a penalty-taker. I have assumed thus far that this specious present has just five phases, P1-P5, but this is just a simplifying assumption: in reality there might be many more – quite how finely our experience can be divided into distinct temporal phases is an as yet unresolved empirical issue. When acknowledging that our experience is divisible into briefer and briefer parts, the Extensionalist is *not* thereby committed to holding that it can be divided into parts that are so brief they are entirely static and motion-free. For as Bergson argued, it is implausible in the extreme to suppose that motion-as-we-experience it can be composed of total immobilities.

Since the Extensional model explains how our consciousness can be as continuous as it seems to be – and as Bergson believed it was – it might seem to provide everything he sought. However, in one respect at least it may well be distinctly un-Bergsonian. One of the recurring themes of *Matter and Memory* is the extent to which memory, in diverse forms, impacts on consciousness. On some interpretations, in addition to distinguishing habit-memory and recollective-memory, Bergson also posited a more primitive and pervasive form of memory: the 'primal memory' (as Barnard calls it) which is responsible for the existence of duration itself. Here is one such passage:

> If ... what you are considering is the concrete present such as it is actually lived by consciousness, we may say that this present consists, in large measure, in the immediate past. In the fraction of a second which covers the briefest possible perception of light, billions of vibrations have taken place ... Your perception however instantaneous, consists then in an incalculable multitude of remembered elements. *Practically we perceive only the past* ... Consciousness, then illumines, at each moment of time, that immediate part of the past, which, impending over the future, seeks to realize and to associate with it.
>
> MM: 193–4

If a primitive form of memory is responsible for the creation of duration itself, then the Bergson's conception of temporal consciousness looks to be very similar to that offered by the *Retentional* theorist. According to the latter, as we saw earlier, a specious present consists of a momentary primal impression and a collection of memory-like reproductions of recent experiences, and it is this combination of ingredients that is responsible for our experience of change.

The account provided by the Extensional theorist is altogether different. When one hears a rapid succession of notes *do-re*, the *do* that is experienced as flowing into the *re* is the *do*-tone itself, the tone that figures in your original perceptual experience. This tone is *not* a memorial reproduction of any kind, it does not seem 'past' in any sort of way – it has all the force and vivacity of any live auditory experience – though it does of course occur earlier than the *re*-tone it flows into. In effect, for the Extensionalist, all the contents of our specious presents are *primal impressions* (as Husserl called them). We are able to experience successions *as* successions because our conscious apprehension can span a brief interval of time.

In expunging memory from the direct experience of change, the Extensionalist is not thereby committed to holding – implausibly – that memory, in different forms and ways, does have an impact on the character of our experience. As far as I can see, much of what Bergson says about the influence of memory on experience is fully compatible with the Extensional model. The only change is the elimination of memory from the explanation of what makes base-level phenomenal flow possible (and real). While it's impossible to know for certain, I like to think that if Bergson had known about the Extensional + overlap option he would have found it an attractive one.[15]

If we do conceive of temporal experience along the lines I have been suggesting, is there anything mysterious or paradoxical about it? Anything which resists description or analysis? Barnard, following Bergson's lead, suggests there is:

> Because the manyness/oneness of our consciousness is unlike any external object in the world, it is not irrational to use paradox as a way of accurately describing our inner world. In fact, it is irrational *not* to do so.... *durée* is, as Bergson emphasizes, that which is 'ever the same and ever changing'. It is always the same in that it us utter continuity, it is always changing because it is a flux of sheer novelty. It is both, and yet actually neither. No descriptive term, or even any clever combination of terms, will ever adequately represent *durée*.
>
> Barnard 2011: 15

I have already suggested that the dynamic features of our experience – such as perceived motion – are in an important sense irreducible (just as Bergson argued) and so primitive. However, there is nothing paradoxical in being a primitive. Nor, as far as I can see, is there anything paradoxical about our consciousness being 'many' in the sense that it contains a diverse range of contents, and also 'one', in that these contents are experienced as unified, either simultaneously or as a succession. It may well be that *phenomenal unity*, the unity within and between experiences that we encounter in our ordinary states and streams of consciousness in its synchronic and diachronic forms, is itself a primitive feature of experience, one that cannot be reductively or informatively explained in other terms – I suspect this is the case.[16] Again, there is nothing mysterious or paradoxical in being irreducible.

Bergson often tells us that duration is *indivisible* – for example in the 'The Perception of Change' he writes:

> It is enough to be convinced once and for all that reality is change, that change is indivisible, and that in an indivisible change the past is one with the present.
>
> CM: 185

If by 'indivisible' Bergson simply means that the relevant conscious episodes are unified in a deep and distinctive way, then he is right, but there is nothing mysterious or paradoxical about this. The left and right sides of your current visual experience are deeply unified by virtue of being *experienced together*, as are your visual and auditory experiences. The successive phases of your consciousness are also experienced together; of course, these phases are also experienced *as* successive, but there is nothing mysterious or paradoxical here either: that is simply what experienced temporality involves.[17]

Bergson's claim than in an experience of change 'the past is one with the present' could easily *sound* paradoxical and would *be* paradoxical if it meant that past experiences occurred simultaneously with present experiences. Self-evidently, an experience or event that is in the past cannot also be present. However, there is a more charitable reading: in experienced change, earlier and later experiences are experienced *as* successive, and hence as unified. But in experienced successions, 'unified' does not entail 'simultaneous' – at least, not if we adopt the Extensional view of the specious present.[18]

Time and temporal experience

We know that in his later works Bergson often took *durée* as the basis for his metaphysical model of the whole of reality. If our experience has the structure of an ongoing flux, so too does the rest of the universe. Knowing this much, however, does not tell us precisely how Bergson conceived the nature of time.

Figure 11.4 depicts some of the main ways that contemporary analytic philosophers conceive time. At the top of the figure is an attempt to depict the 'block universe' or 'eternalism'. In this view, the universe consists of a four-dimensional block of events, one that does not contain a privileged present that is constantly moving into the future. Every event has an equal claim on being 'present' in just the way that every location in space has an equal claim on being counted 'here'. If we can be confident of anything, it is that Bergson firmly rejected the eternalist metaphysic, and he did so from his earliest days. However, this still leaves the other two conceptions of time in play.

At the very bottom of the figure is an attempt to depict *presentism*, the doctrine that only this present moment is real, that the past and future are completely non-existent, and the passage of time consists of one momentary present giving way to the next, then the next. In the centre is the half-way house known as the *growing block* model. According to the latter, the past just as much part of concrete reality as the present – the past is just as the eternalist says – but the universe is dynamic by virtue of the fact that new slices of reality are continually being created, as the block *grows* as new presents

Figure 11.4 The main conceptions of contemporary analytic philosophy of time. © Barry Dainton.

are added. For the growing block theorist, the future is utterly unreal, just as it is for the presentist.

Which of these models (if any) would Bergson have subscribed to?[19] I confess I'm not at all sure – and I'm not sure either whether this is simply due to my comparative lack of the familiar with the Bergsonian *corpus* or not. One thing is clear: Bergsonians who adopt the Extensional model of temporal experience who are also tempted by presentism will have to reject the standard form of this theory, the form which says the cosmos consists of nothing but a single *durationless* slice of reality. Since Extensional specious presents have a non-zero duration – perhaps as much as a second, perhaps rather less – the universe must have at least this much temporal breadth.[20] However, since the vast bulk of the past and future are still condemned to non-existence on this variant of the doctrine, there is no reason for presentists to feel too unhappy with the outcome.

All presentists face a serious – a potentially lethal – problem: how to go about accounting for truths about the past. The Berlin Wall fell in 1989. If the events of 1989 are as real as any current event – as on the eternalist conception of time – there is no difficulty in understanding how statements about 1989 can be true. The situation is otherwise for presentists. How can there be truths about what does not exist? Are claims about past occurrences metaphysically equivalent to claims about events in fictions, such as *The Lord of the Rings* or ancient myths? Perhaps not for statements such as 'The Berlin Wall fell in 1989', since – as of now, at least – there is a vast amount of existing evidence (in people's memories, in newspapers and TV reports) that the Wall did indeed fall in 1989. But what of a claim such as 'There were 50 billion jelly-fish in the Earth's oceans 50 million years ago'? – i.e. claims for which there is no existing evidence. Are statements such as these on a par with 'Hobbits once inhabited The Shire'?

On one (Deleuze-inspired) interpretation, Bergson relied on 'pure memory' to secure the factuality of the past. As events cease to be present they cease to exist, but *that* they occurred – and every detail about their character – is registered and preserved in an 'ontological past' (or 'the past in general'), and in this form the past co-exists with the present.[21] The past thus construed is not a part of concrete reality, it is a *sui generis*, eternal and self-sustaining mode of being – and one whose contents are constantly expanding as new presents are continually created, and new events occur. Needless to say, this doctrine of the past is not easy to make sense of, but presentists who wish to hang on to the factuality of the past are inevitably forced into exotic metaphysics: if ordinary concrete reality is confined to the momentary present, we have to appeal to something *extra*-ordinary to rescue the past from oblivion.

Perhaps Bergsonians needn't take the path of rendering the past and present co-existent. There are a number of considerations which suggest that Bergson himself took the past to be just as real as the present, a stance which is tantamount to endorsing the growing block model of time.

On various occasions, Bergson claims that the duration of the present is attention-dependent. In 'The Perception of Change' he suggests that by appropriately focusing (or relaxing) our attention, we can expand – or contract – the temporal breadth of *durée*.

> My present, at this moment, is the sentence I am pronouncing. But it is so because I want to limit the field of my attention to my sentence. This attention is something that can be made longer or shorter, like the interval between two points of a compass ... an attention which could be extended indefinitely would embrace, along with the preceding sentence, all the anterior phases of the lecture and the events which preceded the lecture, and as large a portion of what we call our past as desired. The distinction we make between our present and past is therefore, if not arbitrary, at least relative to the extent of the field which our attention to life can embrace.
>
> CM: 178–9

On the face of it, extending the experienced present indefinitely far into the past will only be possible if the earlier phases of experience these expansions of the present embrace are themselves real.

With regard to the status of the past it must also be borne in mind that Bergson devotes much of part III of *Matter and Memory* to developing an extended argument that it is difficult to interpret as anything other than an attempt to persuade us that past psychological states are real, even if they are unconscious – in the sense (I take it) of not being among the objects we are currently apprehending. He grants that we find it entirely natural to think that space 'appears to preserve indefinitely the *things* which are there juxtaposed [even when unperceived], while time in its advance devours the *states* which succeed each other in it' (MM: 209). However, Bergson is adamant that this way of thinking is mistaken. During our waking hours our consciousness is largely preoccupied with action: what should we do next, given what we are currently perceiving of our surroundings? Since the vast bulk of the past is irrelevant to current practical concerns, our consciousness simply excludes it: 'On reluctance to admit the integral survival of the past has its origin, then in the very bent of our psychical life – and unfolding of states wherein our interest prompts us to look at that which is unrolling, and not at that which is entirely unrolled' (MM: 219). In reality, suggests Bergson, mental states that are located in the past, and physical objects that go unobserved, are ontologically on a par: both are fully real: 'restore to consciousness its true role: there will be no longer any more reason to say that the past effaces itself as soon as perceived, than there is to suppose that material objects cease to exist when we cease to perceive them' (MM: 2–7).

Admittedly, for much of this discussion Bergson is concerned with the reality of past *memories*, so there is – I concede – a question mark over whether he really intended to extend full reality to *all* kinds of mental state, even if he at times says as much. However, there is one further reason for thinking Bergson was firmly committed to the full reality of the past.

As we saw earlier, Bergson did not shy away from the radical consequences of his direct realist view of perception. When we look at the distant stars, these stars – or their surface features – are directly present to our consciousness, they are literally *in* our consciousness.

> ... even physically, man is far from merely occupying the tiny space allotted to him ... For if our body is matter for our consciousness, it is co-extensive with our

consciousness, it comprises everything we perceive, it reaches as far as the stars....
we are really present in everything we can perceive.

TS: 221–2

So, thanks to our senses, our consciousness extends beyond our bodily confines. Now, if this is the case, then our consciousness must also be able to extend into the distant *past*. After all, the stars we perceive when we look up at the night sky today are the stars as they were hundreds or thousands (or millions or billions) of years ago – even the closest star is more than two light years away, most are much further. Since Bergson was surely well aware of this fact, he clearly saw no difficulty at all in supposing that past events could be directly apprehended – with all their secondary qualities intact – by presently existing conscious subjects. For this to be possible, past objects and events must be just as real as present objects and events. After all, a non-existent star cannot be present in the consciousness of anyone.

So, when it comes to the nature of time, there are several reasons for placing Bergson firmly in the camp of the growing block theorists, and *not* in the camp of the presentists. Bergson may also have been committed to some idiosyncratic doctrines concerning the nature of memory – cf. his claim that we never really forget our earlier experiences. However, if the past is just as real as the present, these memory-related doctrines are not needed to render it possible for there to be *facts* about the past. If I am right, Bergson's conception of the past is less memory-dependent than it has sometimes been assumed to be.

Notes

1. Also relevant are many of the chapters in Skrbina (2009), Alter and Nagasawa (2015) and Bruntrup and Jaskolla (2016); for a useful overview and assessment of recent work see Chalmers (2015).
2. Russell's shift to a version of neutral monism was well underway by the time *The Analysis of Mind* appeared in 1921 but his views underwent several transformations – for a useful overview see Wishon (2015).
3. For more on Bergson's distinctive views on the nature of matter see Sinclair (2020: chapter 5)
4. For more on Bergson's distinctive views on the nature of perception see Watt (2022).
5. The situation is a *little* better in Alter and Nagasawa's *Consciousness in the Physical World: Essays on Russellian Monism* (2015): here Bergson does get one mention, albeit in an excerpt from Russell's *Analysis of Matter* – there is no mention of him at all in the rest of the book, devoted to more recent writings.
6. Similarly, when Russell later made this claim regarding the centrality of *memory* to mentality: 'What characterizes experience is the influence of past occurrences on present reactions ... I think it would be just to say that the most essential characteristic of mind is memory, using this word in its broadest sense ...' (Russell 1956: 153–4).
7. Russell was fond of saying there were two types of philosopher: those who think of the world as a bowl of jelly (i.e. incapable of being cut into parts, and such that if you touch one part you set of vibrations through the rest) and those who think of it as a bucket of shot (i.e. as consisting of discrete independent atoms) – and that a crucial step in his own philosophical development was giving up the jelly for the shot. Given

some of Bergson's formulations in *Matter and Memory* and elsewhere there can be little doubt that Russell (rightly) classed Bergson as being firmly in the 'jelly' camp. For more on Bergson's distinctive holism see Dainton (2022).
8   For more on this topic see Coleman (2014).
9   This is not to say objects appear in the same way to all subjects, or to subjects of different natures or constitutions. Beings with a higher subjective rate of temporal flow will have the capacity to perceive a greater number changes in a given interval (of clock time) than beings running a slower rate. Or, as Bergson himself puts it: '... homogeneous time is ... an idol of language, a fiction whose origin is easy to discover. In reality, there is no one rhythm of duration; it is possible to imagine many different rhythms which, slower or faster, measure the degree of tension or relaxation of a different kind of consciousness and thereby fix their respective places in the scale of being.' (MM: 275)
10 The original New Realism foundered in large part because of certain problem encountered by most direct realists – e.g. handling illusions in a plausible manner. In recent decades, analytic philosophers interested in direct realism have expended considerable efforts in attempts to overcome precisely these difficulties. For an overview see http://plato.stanford.edu/entries/perception-problem/
11 See Ansell Pearson and Mullarkey (2002: 366–7).
12 In Dainton (2022), I discuss Bergson's claims that *durée* is irreducible and indivisible.
13 For more on Stern's views see Dainton (2017a).
14 For more on this problem see Dainton (2014).
15 For further discussion of Bergson and the Extensional view see Dainton (2017b) and Wolf (2021).
16 This is precisely the conclusion I reach in Dainton (2000). In Dainton (2022), I argue that Bergson may well have shared this view.
17 It may well be that by *indivisible* Bergson means something further: that the contents of a single specious present are bound together in such a manner that the contents could not exist independently of one another, or in a specious present with a different overall character. I investigate the issue of 'phenomenal holism' in a more general way in Dainton (2010); in Dainton (2022), I link these investigations to some suggestive remarks of Bergson's.
18 As we saw earlier, for Retentional theorists, contents that are experienced as successive *do* exist in consciousness simultaneously, but the past-seeming contents are 'retentions' of earlier experiences, not the earlier experiences themselves.
19 For some very useful thoughts on this topic also see Fischer (2021) and Moravec (2022).
20 See Dainton (2010b, §6.11) for a version of presentism which is compatible with the Extensional view of temporal experience.
21 I am relying here on M.R. Kelly (2008).

References

Alter, T. and Y. Nagasawa, eds (2015), *Consciousness in the Physical World: Perspectives on Russellian Monism*, Oxford: Oxford University Press.
Ansell Pearson, K. and J. Mullarkey, eds (2002), *Henri Bergson: Key Writings*, New York: Continuum.

Chalmers, D. (1996), *The Conscious Mind*, Oxford: Oxford University Press.
Chalmers, D. (2016), 'Panpsychism and Panprotopsychism' in G. Bruntrup and L. Jaskolla (eds), *Panpsychism: Contemporary Perspectives*, Oxford: Oxford University Press.
Coleman, S. (2014), 'The Real Combination Problem: Panpsychism, Microsubjects, and Emergence', *Erkenntnis*, 79 (1): 19–44.
Barnard, W. (2011), *Living Consciousness: The Metaphysical Vision of Henri Bergson*, New York: SUNY.
Bruntrup, G. and L. Jaskolla, eds (2016), *Panpsychism: Contemporary Perspectives*, Oxford: Oxford University Press.
Chalmers, D (2015), 'Panpsychism and Panprotopyschism' in T. Alter and Y. Nagasawa (eds), *Consciousness in the Physical World: Perspectives on Russellian Monism*, Oxford: Oxford University Press.
Dainton, B. (2000), *Stream of Consciousness*, London: Routledge.
Dainton, B. (2010a), 'Temporal Consciousness', http://plato.stanford.edu/entries/consciousness-temporal/
Dainton, B. (2010b), *Time and Space*, London: Routledge.
Dainton, B. (2010c), 'Phenomenal Holism', *Royal Institute of Philosophy Supplement* 67: 113–39.
Dainton, B. (2014), 'Flows, Repetitions and Symmetries: Replies to Lee and Pelczar', in L.N. Oaklander (ed.), *Debates in the Metaphysics of Time*, 175–212, London: Bloomsbury Academic.
Dainton, B. (2017a), 'William Stern's *Psychische Prasenzzeit*' in I. Phillips (ed.), *The Routledge Handbook of Philosophy of Temporal Experience*, London: Routledge.
Dainton B. (2017b), 'Bergson on Temporal Experience and *Durée Réelle*', in I. Phillips (ed.), *The Routledge Handbook of Philosophy of Temporal Experience*, London: Routledge.
Dainton, B. (2022), 'Irreducibility, Indivisibility and Interpenetration', in M. Sinclair and Y. Wolf (eds), *The Bergsonian Mind*, New York: Routledge.
Dummett, M. (1994), *Origins of Analytic Philosophy*, Cambridge, MA: Harvard University Press.
Fischer, F. (2021), 'Bergsonian Answers to Contemporary Persistence Questions', *Bergsoniana* [on line], available at http://journals.openedition.org/bergsoniana/448; DOI: https://doi.org/10.4000/bergsoniana.448 (accessed 31 March 2022).
Holt, E.B. (1912), 'The Place of Illusory Experience in a Realistic World', in E.B. Holt, W.T. Marvin, M. Pepperel, R.B. Perry, W.B. Pitkin and E.G. Spaulding, *The New Realism: Cooperative Studies*, New York: Macmillan.
James, W. (1890), *The Principles of Psychology*, New York: Henry Holt.
James, W. (1904), 'Does Consciousness Exist?' *Journal of Philosophy, Psychology, and Scientific Methods*, 1 (18): 477–91.
James, W. (1920), *The Letters of William James*, Volume II, Boston: The Atlantic Monthly Press.
Kelly, M.R. (2008), 'Husserl, Deleuzian Bergsonsim and the Sense of the Past in General', *Husserl Studies* 24 (1): 15–30.
Moravec, M. (2022), 'A Bergsonian Response to McTaggart's Paradox' in M. Sinclair and Y. Wolf (eds), *The Bergsonian Mind*, New York: Routledge
Russell, B. (1921), *The Analysis of Mind*, New York: Macmillan.
Russell, B. (1927), *The Analysis of Matter*, London: Routledge.
Russell, B. (1956), *Portraits from Memory*, New York: Simon and Schuster.
Sinclair, M. (2020), *Bergson*, London: Routledge.
Sinclair, M. and Y. Wolf, Y., eds (2022), *The Bergsonian Mind*, New York: Routledge.

Skrbina, D., ed. (2009), *Mind that Abides: Panpsychism in the New Millennium*, Amsterdam: John Benjamins.

Strawson, G. (2006), 'Realistic Monism: Why Physicalism Entails Panpsychism', *Journal of Consciousness Studies*, 13 (10–11), 3–31.

Stern, L. (1897/2005), 'Mental Presence-Time', trans. N. De Warren, in C. Wolfe (ed.), *The New Yearbook for Phenomenology and Phenomenological Research*, 205–16, London: College Publications.

Stubenberg, L. (2016), *Neutral Monism*, https://plato.stanford.edu/entries/neutral-monism/

Watt, R. (2022), 'The Naïve Realism of Henri Bergson', in M. Sinclair and Y. Wolf, (eds), *The Bergsonian Mind*, New York: Routledge.

Wolf, Y. (2021), '"A Memory within Change Itself." Bergson and the Memory Theory of Temporal Experience', *Bergsoniana* [online], http://journals.openedition.org/bergsoniana/286; DOI: https://doi.org/10.4000/bergsoniana.286 (accessed 31 March 2022).

Williamson, T. (2014), 'How Did We Get Here From There? The Transformation of Analytic Philosophy', lecture in Belgrade available online: http://www.philosophy.ox.ac.uk/__data/assets/pdf_file/0006/35835/How_did_we_get_here_from_there.pdf

Wishon, D. (2015), 'Russell on Russellian Monism', in T. Alter and Y. Nagasawa (eds), *Consciousness in the Physical World: Perspectives on Russellian Monism*, Oxford: Oxford University Press.

12

What Arranges Memories in a Line?

Takahiro Isashiki
Nihon University

There is a gap between Bergson's theory of the present and his theory of the past. This gap concerns the concept of 'date' and can be filled with my theory of events. However, in doing so, Bergson's realism about the past must be modified into semi-realism.

1 Three characteristics of Bergson's theory of the past

Bergson's theory of the past has three characteristics: (1) contemporaneous formation of the past and the present, (2) automatic conservation of the past (i.e. past-realism), and (3) trans-temporality of the past[1].

1.1 Contemporaneous formation of the past and the present

Bergson says:

> the formation of memory is never posterior to the formation of perception; it is contemporaneous with it. Step by step, as perception is created, the memory of it is projected beside it, as the shadow falls beside the body.
> ME: 157[130]

> [The present] is twofold at every moment, its very up-rush being in two jets exactly symmetrical [i.e. perception and memory].
> ME: 160[131]

Every time the present appears, it branches into perception and memory. So, memory and perception appear at the same time; perception does not change into memory. In other words, the present does not change into the past. Suppose you heard thunder outside the window right now. The sequence of events is not that first the thunder was perceived, then the perception disappeared, and after that the memory of having heard the thunder appeared. In reality, the thunder was heard and its memory was formed simultaneously. This is the 'contemporaneous formation of the past and the present'.

However, perception and memory have opposite destinies. Perception disappears immediately, although memory does not disappear. Bergson says the following about the disappearance of perception:

> when we add to the present moment those which have preceded it, ... we are not dealing with these moments themselves, since they have vanished for ever
>
> TFW: 79[59]

Certainly, 'new' presents appear successively, so the present itself never disappears. However, perception vanishes just after it appears.

1.2 Automatic conservation of the past

Perceptions disappear quickly, whereas the memory, which appeared together with perception, does not disappear but is stored.

> there will no longer be any more reason to say that the past effaces itself as soon as perceived, than there is to suppose that material objects cease to exist when we cease to perceive them.
>
> MM: 182[157]

When you hear thunder, the perception of it disappears immediately; however, the memory of having heard it remains. Just as material objects do not disappear when you stop perceiving them, the memory of the thunder does not disappear when its perception disappears.

This conservation is neither practiced by human brains nor by human minds. Memories are automatically stored without human engagement.

> the piling up of the past upon the past goes on without relaxation. In reality, the past is preserved by itself, automatically.
>
> CE: 5[5]

> survival of the past *per se*
>
> MM: 193[166]

Even if there were no human beings, the past would be stored. This is the 'automatic conservation of the past', which is the second characteristic of Bergson's theory of the past. He claims that the past continues existing by itself without human intervention. This means that he maintains past-realism.

1.3 Trans-temporality of the past

Since the past continuously piles up, 'the past grows without ceasing' (CE: 5[4]), but then, does the 'new' past change into the 'older past' and become more and more distant

What Arranges Memories in a Line?

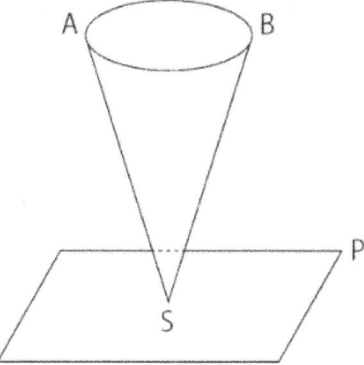

Figure 12.1 Bergson's inverted cone. Henri Bergson. Public domain.

from the present as time goes by? No, it does not. Bergson speaks of an inverted cone, SAB, which symbolizes the entire memory (see Figure 12.1):

> At S is the present perception which I have of my body, that is to say, of a certain sensori-motor equilibrium. Over the surface of the base AB are spread […] my recollections in their totality.
>
> MM: 210[180]

It is not that the more recent memory is near the present (S) nor that the further away from S, the older the memory is. The whole of my various memories is on the same plane, the base AB. Moreover, the base AB is 'motionless' (*immobile*) (MM: 196[169]). Thus, the entire memory (AB) is located at the same distance from the present (S). The 'older' past is not located further away from the present. This is the 'trans-temporality of the past'.

2. The concept of 'date' and Bergson's theory of the present

2.1 The 'date' of past events

Let us look more closely at what Bergson says about the automatic conservation of the past:

> [Spontaneous memory] records, in the form of memory-images, all the events of our daily life as they occur in time; it neglects no detail; it leaves to each fact, to each gesture, its place and date.
>
> MM: 92[86]

that memory which is spontaneous, which dates events and records them but once.

MM: 95[89]

'Spontaneous memory' (*le souvenir spontané, la mémoire spontanée*) is in contrast to 'acquired memory' (*souvenir acquis*). Acquired memory is habitual memory embedded in our body's practical function. One example of acquired memory is learning a passage by heart through repetition (MM: 89–94[83–8]). In the case of acquired memory, we tend to forget when and where we learned it. In contrast, in the case of spontaneous memory, we remember when, where and in what situation we learned it. We recall it as an individual event in our life's history (MM: 90[84]).

Therefore, it is not acquired memory but spontaneous memory that automatically conserves the past. Spontaneous memory gives a date to each event without exception; therefore, it follows that the past has dates. Suppose you heard thunder on 17 September 2015, at 9:48. At the time, the perception of the thunder and the memory of hearing it appeared simultaneously. The perception disappeared at once, but the memory was automatically stored.

However, here is a problem we cannot overlook. It is concerned with the concept of 'date' (*date*). The concept of 'date' can exist only if we presuppose linear time. The 17th of September is after the 16th September and before the 18th September. Here is an order of 'before' and 'after'. Between any two dates there is necessarily an order of 'before' and 'after'. Thus, dates form a linear order. All dates are arranged in a line. Dates are further divided into times, and times are thought to be infinitely identifiable. For example, it is a familiar fact that the times of eclipses and major earthquakes are specified down to the second. After all, dates (and times) assume linear and infinitely divisible time. However, this raises a question.

2.2 Numerical multiplicity and qualitative multiplicity

In *Time and Free Will* (*Essai sur les données immédiates de la conscience*), Bergson criticized the notion of considering time to be linear and divisible.

> We may therefore surmise that time, conceived under the form of a homogeneous medium, is spree spurious concept, due to the trespassing of the idea of space upon the field of pure consciousness.
>
> TFW: 98[73]

'Time, conceived under the form of a homogeneous medium' is linear time, while 'the field of pure consciousness' is real time, that is, 'pure duration' (*durée pure*).

> pure duration might well be nothing but a succession of qualitative changes, which melt into and permeate one another, without precise outlines, without any tendency to externalize themselves in relation to one another, without any affiliation with number
>
> TFW: 104[77]

Bergson sharply distinguishes between numerical multiplicity (*multiplicité numérique*) and qualitative multiplicity (*multiplicité qualitative*). An example of numerical multiplicity is 'counting sheep' (TFW: 76[57]). When we count some things, the counted things are homogeneous (they are all sheep), but they are reciprocally distinguished as individuals. Thus, numerical multiplicity has characteristics of homogeneity (*homogénéité*) and reciprocal exteriority (*extériorité réciproque*). In contrast, an example of qualitative multiplicity is 'listening to a melody' (TFW: 99–104[74–80]). A melody is a qualitative change of sounds. Many sounds that have different pitches, lengths and tones appear one after another and form one complete melody. Each sound that goes into forming a melody penetrates into one another. If we separate them (e.g. if we lengthen a ten-second melody into a five-minute melody and listen to it), they will not be heard as a melody. Qualitative multiplicity has the characteristics of heterogeneity (*hétérogénéité*) and mutual penetration (*pénétration mutuelle*). Bergson says the following:

> if it did not betake itself to a symbolical substitute, our consciousness would never regard time as a homogeneous medium, in which the terms of a succession remain outside one another.
>
> TFW: 124[92]

Thus, linear time, which is numerical multiplicity, is not real time but only a symbolical representation. In contrast, pure duration, which is qualitative multiplicity, is real time.

Then what is the connection between the concept of 'linear time' and that of space? Bergson says that the concept of 'order' presupposes that of space.

> we could not introduce order among terms without first distinguishing them and then comparing the places which they occupy; . . . if we introduce an order in what is successive, the reason is that succession is converted into simultaneity and is projected into space. . . . the idea . . . of a certain order of succession in time, itself implies the representation of space.
>
> TFW: 102[76]

Also, since order is the essence of number, the concept of number presupposes the concept of space as well.

> every clear idea of number implies a visual image in space.
>
> TFW: 79[59]

Therefore, linear time (as numerical multiplicity) presupposes the concept of space, while pure duration (as qualitative multiplicity) never contains the concept of space.

2.3 Separation of 'succession' and 'order'

Bergson admits that time contains 'succession' (*succession*), but he does not admit that it contains 'order' (*ordre*).

> That time implies succession I do not deny. But that succession is first presented to our consciousness, like the distinction of a 'before' and 'after' set side by side, is what I cannot admit. When we listen to a melody we have the purest impression of succession we could possibly have,... and yet it is the very continuity of the melody and the impossibility of breaking it up which make that impression upon us.
>
> CM: 176[166]

He separates 'succession' and 'order'. 'Succession' is qualitative multiplicity, while 'order' is numerical multiplicity. Order needs division into 'before' and 'after', but qualitative multiplicity, the essence of which is mutual penetration, is indivisible, because it would lose its quality if divided.

In short, the concept of order 'before' and 'after', which is presupposed by linear time, needs the concept of space, while pure duration as real time does not contain the concept of space, and so neither does it contain the concept of number. Therefore, pure duration cannot have dates. This follows from Bergson's theory of the present (or pure duration).

However, according to his theory of the past, every memory automatically conserved has a date, and memory and perception are formed contemporaneously. Therefore, there is a gap between his theory of the present and his theory of the past.

3. Bergson's explanation of 'date'

3.1 Automatic conservation of the past and 'date'

Bergson says the following on the date of memories:

> consciousness ... retains the image of the situations through which it has successively travelled, and lays them side by side in the order in which they took place.
>
> MM: 96[89–90]

What we should focus on here is the phrase 'to lay [images] side by side in the order in which they took place'. When we perceive something, new images occur one after another. If our consciousness lays them side by side in the same order as they occur, it would seem a natural consequence that they are arranged in a date order. In fact, Bergson himself says the following on the relationship between the concept of 'date' and arranging images in their order of occurrence:

> [Memory] retains and ranges alongside of each other all our states in the order in which they occur, leaving to each fact its place and consequently marking its date.
>
> MM: 195[168]

Do his arguments about the automatic conservation of the past really entail that the past has dates? There are three reasons why we doubt it.

First, according to Bergson, the formation of perception does not have 'order'. It has qualitative change and succession but it does not have order. As I said in 2.3, Bergson distinguishes 'succession' from 'order'. Qualitative multiplicity has 'succession' but does not have 'order'. For him, 'order' is a spurious concept which can exist only after the concept of space intrudes into pure duration. Since the formation of perception does not have order and memories are formed at the same time as perceptions, we should say that memories cannot have order (or dates).

Secondly, even if we assume that the formation of memory could have 'order', the phrase 'to lay images side by side in the order in which they took place' is ambiguous. It can mean (1) making sure that the action of arranging is done in a definite order (i.e. the order of occurrence of memories) and (2) making sure that the result of arranging is in a definite order (i.e. the same order as (1)). What is important is that the former (1) does not imply the latter (2). Metaphorically, it is similar to the ambiguity of the phrase 'to seat people side by side in the order in which they arrive at the hall' (see Figure 12.2). It can mean (3) to let people enter the hall and sit on a seat without making them wait or (4) to make the order of their seats correspond to the order of their arriving at the hall. The former (3) means making sure that the two temporal orders (i.e. the order of arriving and that of sitting) are the same. The latter (4) means making sure that the temporal order (i.e. the order of arriving) and the spatial order (i.e. the order of seats) are the same. Even if we sit immediately after we enter the hall, we can sit on any seat (unless the seats are reserved). In this case, (3) is true but (4) is false; therefore, (3) does not imply (4). Likewise, (1) does not imply (2). The order of the actions does not always determine the order of the results. After all, even if there were an order in the formation of memory, this does not imply that the memory content itself has an order (i.e. a date).

Thirdly, if we assume linear time, it should follow that memory has date, but Bergson does not admit linear time. If we assume that time is linear, we can obtain three propositions: (a) time spreads out like space; (b) the future is real; and (c) the future changes through the present into the past. Thus, events arranged in the future time would move into memory in their order and consequently memory would have dates (see Figure 12.3).

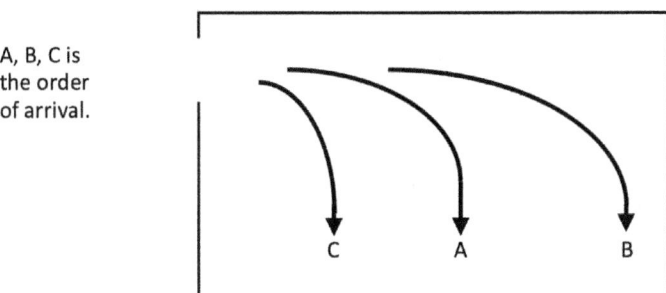

Figure 12.2 The order of arrival. © Takahiro Isashiki.

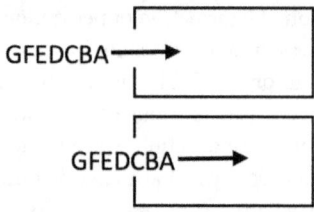

Figure 12.3 The order of events in linear time. © Takahiro Isashiki.

However, Bergson admits none of these three propositions. As I said in section 2, for him, real time is pure duration and it contains no elements of space. Thus, he denies (a). Also, according to him, the future's essence is its unpredictability.

> there is unceasingly being created in [time] ... something unforeseeable and new.
> CE: 359[339]

The essence of time is that it constantly creates something unpredictable and new. Therefore, the future is not real. Thus, he denies (b). Finally, as I said in section 1, perception and memory are formed contemporaneously, and so the present does not change into the past. Thus, he denies (c).

For these three reasons, we can conclude that the automatic conservation of the past does not entail that the past has dates.

3.2 Unrepeatability of the past and 'date'

In fact, Bergson explicitly states the reason why he thinks memory has dates. It is the connection between 'date' and the unrepeatability of the past.

> the immense majority of our memories bear upon events and details of our life of which the essence is to have a date, and consequently to be incapable of being repeated.
> MM: 94[88]

Bergson asserts that the past is unrepeatable because it has dates. Certainly, if an event has a date, it is impossible for it to be repeated. 'To be repeated' means 'to occur again at a different time'. Thus, if events necessarily have dates, an event with a different date must be a different event and consequently the same event cannot be repeated. Certainly, having dates implies unrepeatability. Thus, dates are a sufficient condition of unrepeatability. In this respect, Bergson is right.

However, are dates a necessary condition for unrepeatability? Being a sufficient condition does not always mean being a necessary condition. In reality, dates are not a necessary condition for unrepeatability, as I will show below.

What is necessary for unrepeatability of events is not their dates but their individuality. Suppose you hear two crashes of thunder successively outside the window. These crashes are individual events. Two crashes of thunder rolling successively do not imply a repetition of the past. The reason events are individual is that they occur only once. So, they are unrepeatable, whether or not they are arranged in a line. The concept of order (or date) is not necessary for unrepeatability of events.

Against this, someone may argue that order is necessary for events to have individuality itself. He or she may say that without the distinction between the 'former' crash and the 'latter' crash, there cannot be different crashes, and so the order of 'before' and 'after' is necessary for the very individuality of events.

However, according to Bergson, those crashes of thunder are not distinguished by their order such as 'before' and 'after'. They are distinguished by their qualities. The one is a crash containing a foretaste of the other, and the other is a crash containing the remains of the one; because of these differences in quality, they are different events. Until the concept of space intrudes, we cannot externalize or count these events as the 'former' crash and the 'latter' crash.

If you want to read 'order' in the 'foretaste' and 'remains' themselves, it is due to your deep-seated tendency to understand time through the concept of space. This is similar to the mistake of thinking that a sheet music is essential to understand a melody. We can understand a melody even if we do not replace the succession of sounds by a spatial order, namely a musical notation. Certainly, after hearing a melody, we can write it down in musical notation. However, it is a confusion of qualitative multiplicity and numerical multiplicity to think that the melody originally contained these elements just because they can be broken down into elements after the fact. Similarly, qualitative multiplicity is sufficient for individuality of events; numerical multiplicity such as 'order' is not necessary for it. If we follow Bergson's theory of the present, we should conclude this.

Probably Bergson uses the word 'date' because he wants to say that memories of events are individual. Certainly, events with dates are individual; that is, events having order and arranged in a line are individual, but being arranged in a line is not necessary for the individuality of events. This is because their individuality is secured by their qualitative multiplicity. Let us call such individuality 'qualitative individuality'. Successive crashes of thunder are individuated by their qualities. Therefore, even if events do not have order, they are unrepeatable as long as they have qualitative individuality.

In short, dates are not a necessary condition for the unrepeatability of the past. Therefore, the reason why the past has dates is not clearly explained in Bergson's theory of the past. What, then, are dates derived from?

4. What kind of entity are events?

Dates are ascribed to events. So, what kind of entity is an event? Let us put Bergson aside for a moment and consider the entity of the event.[2]

4.1 The category of events and things

Look at the world around us. There are many things, and many events occur. Some things change their properties. For example, a green leaf turns red in autumn. It is the same leaf even if its colour changes. The concept of 'change' is possible only if there is a thing that retains its numerical identity through the change (i.e. the subject of change).

We can distinguish between thing-individuals and thing-types. For example, in a school classroom, there are several desks, that is, there are several thing-individuals belonging to the same thing-type of 'desk'. Similarly, we can distinguish event-individuals and event-types. For example, a 'wedding' is an event-type and 'my wedding held about forty years ago' is an event-individual.

Regarding events, we can distinguish between 'events already finished' and 'events not yet finished'. As far as events already finished are concerned, no contradictory descriptions are allowed for any detail. For example, how many pictures were taken at my wedding about forty years ago? We cannot count it exactly now, but we must believe that the number is definite. It cannot be both more than a hundred and less than a hundred; that is a contradiction. In contrast, for events not yet finished, contradictory descriptions are allowed. For example, at my son's wedding, which will be held next Sunday, the number of pictures taken may be more than a hundred or less than a hundred. It is irrelevant to the identity of the wedding. (In either case, it will be 'my son's wedding'.)

Even after an event begins, there are various possibilities with regard to its details until it is finished. For example, the number of pictures taken during my son's wedding will not be definite until it is finished. Thus, it is also an 'event not yet finished'.

In short, events not yet finished can be described in various ways, even in contradictory ways, and so they are not event-individuals but event-types. However, once they are finished, contradictory descriptions are no longer allowed and only a definite description is permitted. At this moment, an event-individual appears. For example, at the moment my son's wedding has finished, an event-individual called 'my son's wedding' will appear. Therefore, only events already finished, that is, what is called a 'past event' can be an event-individual.

It follows that thing-individuals and event-individuals have different characteristics with regard to their 'start' and 'end'. The 'start' and 'end' of a thing-individual (e.g. a human being's birth and death) are the start and end of its existence. In contrast, the 'start' and 'end' of an event-individual (e. g. the opening and closing of my son's wedding) are only the start and end of its becoming. The moment the process of its becoming completes, it begins to exist.

Event-individuals and thing-individuals exist in different ways. They belong to different categories: the category of events and the category of things. They are also different with regard to the concepts of 'change', 'disappearance' and 'resurrection'. It is conceptually possible for thing-individuals to 'change', 'disappear' and 'resurrect'. For example, a green leaf turns red (i.e. changes) and then dies off (i.e. disappears). Jesus Christ died on the cross and resurrected after three days. I do not know whether his resurrection was the case but at least it is conceptually possible. On the other hand, it is conceptually impossible for event-individuals to 'change', 'disappear' or 'resurrect'. For example, no matter how similar two events are, we do not say that the same event resurrected. We sometimes change our evaluation of a historical event, but we do not say that by doing so we have changed the event itself. Event-individuals are impossible to change, so they cannot disappear, because disappearance is the limit of change.

Against this, someone may argue that finished events do not exist any longer, that is, they have disappeared; if so, when did they disappear? As I said above, an event-individual appears the moment it is finished. If it disappears when it is finished, then it disappears the moment it appears. That is absurd. And after its appearance there is no opportunity for it to change or disappear. Once an event-individual begins to exist, it never 'changes' or 'disappears'. Event-individuals form networks with other event-individuals and they exist as long as the networks exist.

4.2 Networks of event-individuals

Event-individuals and thing-individuals have some similarities. Both can be referred to, counted and described in multiple ways. They are referable, countable, multi-describable.

Facts are different from event-individuals because they cannot be described in multiple ways. For example, that Caesar was killed and that Caesar was killed by Brutus are different facts; you may know the former but not the latter, but both are facts about one and the same event, which occurred about 2,000 years ago. When we examine one event, we can find several unknown facts about it and describe it in more detail.

However, sometimes the description is not considered to be the same event's new description but rather to be a description about a different event that occupies the same spatiotemporal space with the original event. Suppose the temperature around a rotating ball is increasing, we examine it, and we find that the ball is generating heat. 'The generation of heat' may not be thought to be a new fact about 'the rotating of the ball' but to be a different event which occupies the same place as 'the rotating of the ball'. The boundary between the same event's new description and the description of a new event that occupies the same place is ambiguous.

In addition, among event-individuals, there are relationships such as 'cause and effect' or 'an evaluation of an event'. For example, 'leaving a lighted cigarette unattended' and 'an outbreak of a forest fire' can be cause and effect. 'Running 1,000 meters at a certain speed' and 'winning the race' can be an event and its evaluation.

Understanding an event-type usually includes understanding its relationships with other event-types. For example, when we understand the event-type of 'leaving a lighted cigarette unattended,' we also understand that this is often the cause of fires.

In this way we human beings connect event-individuals by various relationships (like multi-description, occupying the same place, causation and evaluation) and form many event-individual networks. Relying on these networks, we refer to event-individuals, but at the same time, through our referring to event-individuals, their networks are formed and maintained.

Networks become integrated and form larger networks, finally creating a network as the most sophisticated means of reference to event-individuals, namely, 'dates'. This is the origin of 'date'.[3] In short, the 'date' is a network of event-individuals arranged in one line.

4.3 Arbitrariness of individuation of events

Compared with thing-individuals, event-individuals are individuated in more arbitrary ways. First, the same event-individual can be divided into smaller sub-event-individuals, corresponding to the way people refer to them. For example, the event-individual 'London Olympics' contains many small event-individuals: the opening ceremony and many races, and we can refer to even one action of a runner in a race as an event-individual. 'His looking back just before the goal caused other runners to fall down.' In this sentence 'his looking back' and 'other runners falling down' are thought to be an event-individual. One is thought to be a cause and the other its effect.

Secondly, there are also cases in which two event-individuals overlap. If an athlete said, 'It took me three years to win the gold medal at the London Olympics', this event-individual winning the gold medal spans three years from the start of preparation to the moment of victory. The moment was several days before the closing ceremony of the London Olympics. Therefore, this event-individual partially overlaps with another event-individual, the London Olympics.

There is no such arbitrariness in the individuation of thing-individuals. The individuation of events is more arbitrary than that of thing-individuals.

5 Semi-realism about the past

Based on the arguments above, let us examine Bergson's concept of 'date' again. He claims that the individuation of thing-individuals is arbitrary.

> All division of matter into independent bodies with absolutely determined outlines is an artificial division.
>
> MM: 259[220]

In the immediate data of the consciousness, the world is a 'moving continuity'. Its aspect changes at every moment, that is, 'the whole has changed, as with the turning of a kaleidoscope' (MM: 260[221]). However, we human beings dissociate 'permanence' and 'change' from the whole, representing the former by 'bodies' and the latter by 'homogeneous movements in space'. This division comes from the necessity of our lives. According to Bergson, there are no such things as 'independent bodies with an absolutely determined outline' in the world. The entire world is always changing, but human beings extract parts from it that are changing relatively less and see them as 'bodies' i.e. thing-individuals. We do so in order to survive, and the distinction between 'permanence' and 'change' depends on the size of the human body to some extent. If the human body were the size of an ant, what is considered a 'thing-individual' would be completely different from what it is today.

As I showed in 4.3, the individuation of event-individuals is more arbitrary than that of thing-individuals. Therefore, Bergson should say the same thing about event-individuals as about thing-individuals: 'All division of changes of the whole world into event-individuals is an artificial division'. Every change appears and vanishes immediately. We human beings make it 'an event', i.e. an independent entity with an absolutely determined outline. We do so in order to survive. If human beings had extraordinary longevity and had an insensitive sense of time, in which a thousand years would be felt as if it were only a second, what is considered an 'event-individual' would be completely different from what it is today.

When several crashes of thunder roll, they form one qualitative change as a whole. They form a qualitative multiplicity. When they are conserved in memory, their individuality is just that of elements that penetrate into one another in a qualitative multiplicity, that is, qualitative individuality. It is the same as the individuality of sounds in a melody. As I showed in 3.2, elements in a qualitative multiplicity are distinguished by their qualities, not by their 'order'.

However, importing the concept of space, we human beings divide one qualitative multiplicity into event-individuals with a determined outline and change it into a numerical multiplicity by arranging them in an 'order', such as 'the former crash' and 'the latter crash'. This is where 'order' appears. As I showed in 4.2, finally, the 'date', which is a network of event-individuals arranged in a line, appears.

Therefore, to the question 'What arranges memories in a line?' we can answer, 'Human beings arrange memories in a line by creating the category of events'. As a matter of fact, Bergson says in 'Intellectual Effort' published six years after *Matter and Memory*:

> the effort of recall consists in converting a schematic idea, whose elements interpenetrate, into an imaged idea, the parts of which are juxtaposed.
> ME: 203[167]

I think that this comment contains the idea that the automatically conserved elements of memory do not have 'order' by themselves, but when they are recalled they come to have 'order'.

Conclusion

The gap between Bergson's theory of the present and his theory of the past can be filled with my theory of events. Bergson says that the past is automatically conserved whether there are human beings or not, but my arguments have shown that the automatically conserved past has qualitative individuality but not 'order' (or 'dates').[4] The 'order' is given by human beings. Human beings divide one qualitative multiplicity into event-individuals, change it into a numerical multiplicity, connect event-individuals by various relationships, form event-individual networks, and finally create a network as the most sophisticated means of reference to event-individuals, namely, dates. In short, human beings arrange memories side by side in a line by creating the category of events. The past belongs to the world, in that it is automatically conserved, but its 'order' comes from human beings. The past depends partly on the world and partly on human beings. Therefore, Bergson's realism about the past should be modified into semi-realism.[5]

Notes

1. I learned these characteristics from studies by Prof. Yasushi Hirai (Keio University): (2011a, 2011b).
2. I have explained my theory of events in Isashiki (2010).
3. A.N. Prior, the founder of tense logic, says, 'Dates, like classes, are a wonderfully and tremendously useful invention, but they are an invention' (Craig 2000: 246).
4. Events which are individuated but not arranged in a line is an imperfect concept of an event. When we observe the concept of events that little children have, we can know that such an imperfect concept is possible. Two-year-old children have scripts of events (i.e. knowledge of typical relationships among event-types) but they cannot remember the connections among event-individuals. Three-year-old children cannot infer where the toy is when they watch a video showing that the toy was put in a box behind them two minutes ago. Four-year-old children can judge which event-individual is older but they cannot know the season or date when the event occurred. These children have imperfect concepts of events. I have explained my view in Isashiki (2017).
5. Depending on the character of event-individuals' existence, we may be able to thoroughly give up the realism of the past. But this possibility should be investigated as a separate question, rather than in an interpretation of Bergson.

References

Craig, W.L. (2000), *The Tensed Theory of Time: A Critical Examination*, Dordrecht: Kluwer Academic Publishers.

Hirai, Y. (2011a), 'What is Time for Freedom: Bergson on the Freedom without Possibilities', *Annual Review of the Philosophical Association of Western Japan*, 19: 161–86.

Hirai, Y. (2011b), 'The Anatomy of Bergson's Time-Ontology through McTaggartian Problematic', the 62nd Conference of the Philosophical Association of Western Japan, 26 November 2011.

Isashiki, T. (2010), *Metaphysics of Temporal Modality: What is the Present, the Past, and the Future*, Tokyo: Keiso-shobo.

Isashiki, T. (2017), 'Origin of Time Consciousness', in Yukihiro Nobuhara (ed.), *Time, Self, and Narrative*, Shunjusha: 109–42.

13

Coexistence and the Flow of Time

Elie During
University Paris Nanterre

'Time' and 'duration'

Bergson, as is well known, criticized the illusion whereby time is collapsed with its relative measures in the form of 'mathematical time', and as a result loses touch with lived duration, which alone confers its distinctively temporal character on 'time'. However, redefining real time in terms of duration, while leaving aside measured time as a mere spatialized residue, is only half of the story, and probably not the most interesting part. For once the criticism of spatialized time has been carried out, it is equally important to understand how heterogeneous durations unfolding together across the universe can be woven into the fabric of an evolving, open 'Whole' (*Tout*). Here time comes into play once again, not, however, as another name for space or as the universal form of becoming epitomized by the popular image of the cinematograph, but as a *medium of coexistence* to be recovered beneath spatial perspectives and measured durations. Accordingly, the crucial concept here is that of *simultaneity*. Two distant events, separated in space, can be simultaneous if they occur 'at the same time' under some relevant perspective. The challenge for Bergson is to understand that fact temporally: to be simultaneous is a matter of *enduring together in time*, not merely being apart in space.

Such is the crucial problem of the 1922 essay on Einstein's theory of relativity. The title speaks for itself: *Duration and Simultaneity*. The two concepts are not only contrasted, as they generally were in previous works. They are introduced as two interlocking aspects of time. As Kant already acknowledged, *pace* Leibniz, there is more at stake in simultaneity than mere non-succession. If there is a lesson to be learnt from the 'third analogy of experience' in the first *Critique*, it is that simultaneity is a genuine and *sui generis* temporal relation, not the mere obverse of succession. Bergson goes one step further still in suggesting that simultaneity may be as fundamental as succession itself, if not the primitive temporal concept, by contrast with Kant's claim that it is 'one of the most interesting ideas derived from time' (1770 *Dissertation*, section III, §14). To realize this fully, we must embrace a dynamic understanding of simultaneity, predicated on the intuition of contemporaneous *flows* rather than the Kantian community or interaction of substances. Hence the recurring motive of the 'simultaneity of flows', contrasted with the 'simultaneity of instants' (DS: 52). However, unless 'flow' is a verbal garment for the homogeneous, linear, space-like dimension of successive events, it needs to be understood from a durational perspective, i.e. as a vector of real change. This is where *Duration and Simultaneity* has a surprise in store for us. For the flow of duration itself, as it turns out, is ultimately dependent on the coexistence of many durations unfolding in parallel. If the diffuse presence of the universe as a whole did not press upon each of them, time would never 'flow'; it would indeed reduce to an order of simultaneities and succession.

This theme was forcefully introduced in the first chapter of *Creative Evolution*: it is essential to keep it in mind when reflecting on the unity of time in the context of relativity theory, otherwise the fundamental issues quickly get blurred in the tiresome discussions triggered by overly sophisticated arguments about the proper interpretation of 'mathematical' and 'physical' times, their perspectival or non-perspectival character, and the like. Bergson's challenge is to recover an adequate picture of the propension of duration to extend beyond the 'here-now' – that is, not only beyond the now, but beyond the here as well. Commentators have put so much emphasis on the link between time and measurement, rehashing the textbook opposition between 'clock time' and 'conscious time', that this critical involvement of duration with *extension* – irreducible to space as mere quantitative multiplicity – has been almost obliterated.

The criticism of time initiated in *Time and Free Will* was only a prelude to the reconstruction of its concept along philosophical lines, a reconstruction that called for an active confrontation with the scientific representations of becoming, particularly those which, like relativity theory, suggested a radical overhaul of the categorial framework holding for such things as the time elapsed between two successive events; but what is the critical function of duration in that respect? In what way does the doctrine of duration positively contribute to a refoundation of the philosophy of time? Is duration a mere placeholder for consciousness (a reminder that so-called 'subjective time' should not be left out of the picture)? Forget the somewhat sterile opposition of lived time and clock time for a second. Consider, rather, what the concept (or image) of duration achieves in Bergson's overall philosophical strategy. The answer is quite straightforward. Subjective time owes its importance to the fact that it is the most immediate access we can find to a special kind of multiplicity: a purely qualitative, non-serial multiplicity that is best characterized, topologically, by the 'reciprocal penetration' of its parts. More profoundly, as Deleuze's commentaries emphasized, duration shifts the focus to the most fundamental operation of time, namely: the ontological split between present and past in terms of actual and virtual. I shall devote the first half of this chapter to this particular problem, if only to underline Bergson's singular position with respect to the presentism/eternalism debate that rages in the analytic philosophy of time.

From there, I will move on to the topic of coexistence, which I will deal with in a more programmatic way. How do we make sense of the simultaneity of parallel unfolding of durations, not only in experience, from the situated perspective of a living consciousness, but formally, as a metaphysical issue concerning virtually all beings? Before addressing this question, it is important to realize that Bergson, for one, was convinced that to relativize temporal concepts, time itself must first be characterized by its essential unity on a formal level. Duration, on the other hand, necessarily comes in the plural. In fact, it may be argued, based on Bergson's famous example of the sugar block melting in water (CE: 12–13, 368–9), that our best phenomenological access to duration comes in the form of a differential relation between two contrasting rhythms of duration. Thus, the subjective rhythm of my own impatience reflects an incompressible duration, or rather a mesh of entangled durations, as it is pitched against the mineral temporality of the sugar's melting. Although Bergson sometimes attempts a general definition of the concept of duration, emphasizing – as we shall see – a twofold principle of conservation and variation, it would be misleading to assume a single, all-embracing Duration to which all durations could ultimately be referred, performing the function of an ontological principle, a temporal substance of sorts. Even cosmic duration is in the end but one duration among others – the duration of that particular object known as 'the universe'. It is the function of 'time' – that is, the concept of time – to achieve a precise formulation of the associated cosmological problem, namely the problem of the unification of material durations, what *Duration and Simultaneity* refers to as 'the unity of real time'. This concern was already apparent in the last chapter of *Matter and Memory*. To describe matter itself in temporal terms, Bergson referred it to a single ontological plane corresponding to the most relaxed, distended or extended state of duration. There is such a thing as enduring matter, as expressed by the fact that every real motion involves a qualitative change somewhere in the whole to which it belongs. However, to paraphrase Deleuze, this duration exhibits only the lowest degree of difference. Bergson was relying on such assumptions and schemes when he set out to confront relativistic views. The question of the unity of time in a relativistic setting cannot be separated from the underlying issue of the metaphysical unity of matter, conceived as the plane of furthest extension of duration.

The form of time: a Kantian interlude

To anyone who is familiar with the history of the time concept in philosophical discourse, the idea that time should not be conflated with duration hardly comes as a surprise. It is a claim that Kant, for instance, would have readily endorsed. In fact, it is a recurrent theme of the first *Critique*. Time is not a concept but a *form*; not a sensible datum but a 'pure intuition'. We are familiar with such statements. Maybe less so with their immediate consequence, namely that time does not pass, nor flow in any rigorous sense. 'Time does not flow, while everything flows within it' (A 144). Even the pure form of succession is in itself immutable: time, Kant writes, is 'what remains and does not change' (B 225). No wonder that substance (or permanence), the first determination of time according to the categories of relation, plays such a critical role in Kant's doctrine of time. It all comes down to this: time itself – the form of time, time as form – must not be confused with the temporal phenomena or processes unfolding *in* time. Aristotle, it might be remembered, already warned us in the *Physics* that time itself, which allows motion to be measured, does not accelerate or slow down. Time as a transcendental condition of change is not a special kind of duration, more general or abstract than any particular sensible process. It is in fact nothing like a duration. Its position in philosophical discourse is reminiscent of Wittgenstein's idea of 'formal concepts' in the *Tractatus*. Even as we acknowledge that time does not flow, we should not mean it in the sense that it stands still. What we should mean, rather, is that time is not the kind of thing that may flow or not. It is a matter of philosophical grammar. In the same way, to say that time is not made of instants doesn't mean that it is made of finitely small stretches of duration. Whitehead's doctrine of 'atomic' or 'epochal' time, it must be remembered, precedes any particular decision regarding the measurement of extension within the spacetime continuum. Coming back to Kant, time is a form in two main respects: first, it is a formal intuition which is required to make sense of the continuity and connectedness of experience; second, it is a concept that is elaborated and given objective meaning in relation to the postulates of empirical thought, as well as very general connecting principles such as: permanence, succession, and simultaneity, to name those listed in the 'analogies of experience'. Kant's suggestion is that these principles operate through the temporal schematization of the three categories of relation: substance (permanence), causality (succession), community or reciprocal action (simultaneity). The relevant section in the *Critique of Pure Reason* starts in A 176/B 218. Meditating these passages should be a prerequisite for any serious engagement with the Bergson-Einstein dispute.

Thus, time cannot be treated along the same lines as temporal phenomena, or concrete durations. There is a general lesson to be taken here regarding the passage of time. Despite the sophisticated ontological frameworks designed to make sense of the elusive 'now', we are still struggling with all too familiar but ultimately misleading metaphors such as the river of time, the moving spotlight, the growing block and the like. Unpacking their underlying logical structure does not render them any more adequate. The phenomenological approach to time-consciousness has brought its own load of metaphors in connection with the so-called stream of consciousness. In this respect, there is no denying that Bergson's emphasis on the need to rejuvenate our understanding of time by appealing to concrete acts of intuition often seems to suggest that talk about time *in general* must give way to a metaphysical celebration of the multifarious flows of durations. This view, however, must be resisted. The time/duration polarity remains an essential tool for philosophical analysis, and as we shall see, it is the only way we can give a precise formulation to the cosmological issue of the coexistence of concrete durations.

As for the alleged 'flow' of time (or duration, for that matter), a flow that can be speeded up or slowed down, it has more than once proved to be an epistemological obstacle. Physicists know very well that they speak somewhat loosely of a 'slowing down' of time caused by kinematic or gravitational factors. Since everyone seems to cling to the metaphor of passage – even to expose it as unreal – the safest thing to say is that time flows equably, in Newton's parlance, at the uniform rate of *one second per second*. The rhythmic unity of matter comes down to this trivial truth, and Paul Langevin's suggestion that Bergsonian 'real time' could be identified with the physicist's 'proper time' is, under the most charitable interpretation, merely making the same point. What is sometimes described in relativity theory as a 'dilation' of time is only a fancy way of describing discrepancies in measured elapsed durations which are best expressed using intrinsic magnitudes such as proper times. Local measures of time keep desynchronizing, there are delays and dephasings, as illustrated by the famous 'twin paradox', because 'proper time' turns out to be a local, path-dependent magnitude. Gaining speed, relative to others and the surrounding universe, amounts to taking shortcuts in spacetime. There is no need to picture a mysterious temporal substance stretching or dilating. Such manner of talking is positively medieval and should be banned from popular introductions to relativity theory. Nothing happens to time or duration as a result of relative speed or acceleration, but the intervals between certain events admit various measures of time, depending on the way they are causally connected. This fact can be given a consistent interpretation in Einstein-Minkowski's spacetime framework. The philosophical task is to provide a harmonious account of seemingly conflicting figures of time – such as parameter (or 'proper') time and coordinate time, local time and global time – while doing justice to the natural expectations attached to the idea of time. In particular, one should spend some effort finding the most inclusive way of accounting for such basic facts as the persistence of individuals through time, or the coexistence of contemporaneous events and processes across space, without dismissing them wholesale on the grounds that physical theory has proven time and simultaneity to be relative in some sense.

As far as the philosophy of time is concerned, the moral of the situation is quite simple. To repeat, time is not itself a process, but the form of every process. As such, it must be treated as a formal concept (see During 2021). What this means, practically, is that 'time' is essentially the name of a problem, or a cluster of problems, rather than a general denomination subsuming a class of particular objects (whether 'durations' or 'changes'). As a result, any particular figure of time calls for philosophical interpretation, rather than direct application. Interpretation must be carried out according to the 'demands' that the form of time makes on us. Following Kant's characterization of the order of time in the 'analogies of experience', and notwithstanding the Bergsonian shift of focus on flows (or processes) rather than substances, we may say the 'horizon of expectation' associated with time involves three basic coordinates:

1° *Duration* (permanence);
2° *Order* (succession);
3° *Coexistence* (simultaneity).

1° and 2° are typical ingredients of the view of time as the dimension of change. This does not need too much emphasis. A critical question, on this chapter, is that of the type of continuity we wish to grant time in order to account for the unfolding of actual processes. Equally important is the third aspect of time, described under the heading of coexistence: time as a sheaf or envelope of becoming, extending over space, holding together bundles of durations that may flow together, overlap or coexist at a distance, according to simultaneity relations that extend beyond the here-now of coincident events. 2° and 3°, taken together, raise the issue of time's relation with the causal order as a whole.

My contention is that we cannot pick and choose. These three profiles of time are inseparable and come as a package. More than a relational concept, time is a genuine structure of thought in that regard. If a philosophy of time does not bother to address the whole bundle of related, intertwining issues in some way or other, it is not worth one hour's effort – 'elle ne vaut pas une heure de peine', as Blaise Pascal would say. The first philosophical task awaiting us consists of weaving the different dimensions of time together – duration, order and coexistence – rather than reviewing them and analysing them in turn only to leave them in their original state of dispersion. So much for the Kantian interlude. Let this be a blueprint for any future philosophy of time.

Beyond presentism and eternalism

Back to Bergson. For the author of *Creative Evolution*, persistence and change, continuity and emergence – 'the continuous production of endless novelty' – are of course constant metaphysical concerns. Give them up and the whole discussion about the 'passage' of time appears purely verbal and somewhat unreal. Truth be told, debates between tensers and detensers, presentists and eternalists, endurantists and perdurantists always strike Bergsonians as a rather frivolous affair. The suspicion is that the disagreement between the contenders is not really about time, as much as it is about the grammar of existence or identity. As the debate is usually framed, it is clear that the opposing parties differ on such technicalities as the exact scope of the existential quantifier, for instance. Presentists, as they are often portrayed, hold that only what is present exists, while eternalists believe that everything (past, present or future) exists, which obviously cannot be taken as meaning 'exists *now*', unless 'now' ranges over all possible times, past, present or future (to the extent that they were, are or will be 'now'...).

These debates can be widely entertaining if one is in the mood. They involve quirky questions, such as: 'How do we know it is now *now*?' However, to repeat, are we really arguing about *time*? What relevant aspect of time do such debate address, and to what end? If there is certainly much to be learnt there about the grammar of 'exists', or matters of tense and persistence and their connection with questions of identity across possible worlds, it may be argued that the more fundamental issue of *realization* is generally overlooked, or simply taken for granted. Realization does not reduce to the mere 'happening' of things, to the 'formality of taking place', to quote from Eddington. If it were the case, the world as a whole could be recapitulated with each new moment, as if it were created anew. The past could be condensed in the immediately preceding moment, the one that supposedly determines the present. Such a conception may be convenient for practical purpose, but it is metaphysically dubious. As Bergson put it: 'The world the mathematician deals with is a world that dies and is reborn at every instant – the world which Descartes was thinking of when he spoke about continuous creation' (CE: 23–4).

Metaphysicians are indeed often closer to mathematicians than they would like to believe. Think of the way both presentists and eternalists deal with past and present moments: from an ontological point of view, each moment, it seems, can be considered *apart* from all others, as if it were alone, or the very first in the temporal series. The only sense in which the past can be said to continue itself in the present is by means of actual memories, retentions or material traces of past events.[1] All there *is* (now) is supposed to be given in actuality, laid out in full view, including what remains from the past – anything, that is, but the past moments themselves, which are irremediably gone, subsisting only in eternity, as part of the overall order of time. However, the world of the mathematician is not the world we live in, it is not a universe that endures, infused with buzzing life. Nor is it exactly the physicist's world, by the way. For even matter, compared by Bergson to 'a present that is always beginning again' (MM: 178), can only be equated with pure repetition if we suppose it to be entirely extended in space, which Bergson acknowledges is only an ideal limit suggested by the general direction of material process. Even in the limited realm of matter, presentism remains a fiction: it has only tangential validity, it can only be true at the limit when, by a leap of the understanding, matter is identified with instantaneous space.

Laplace's ghost

One is reminded here of Laplace, who derived his ideal of determinism from the assumption that everything there will be, in the future, can be strictly derived from what there is in the present moment, where it pre-exists in compressed form. To apply deterministic laws, becoming needs to be spelled out in terms of successive instantaneous 'nows'. It matters little whether we grant them physical reality or define them as the limits of finite intervals of time. Pictured as infinitesimally thin layers stringed together by external relations of succession, each now appears entirely determined by its immediate past. At every stage the past as a whole appears recapitulated or projected unto the immediately preceding moment – that is, a moment that can be taken as close as one wishes to the 'now' elected as a temporal index. According to this conception, nothing would be changed if the world were destroyed and created afresh at every moment, as Descartes supposed, or infinitely speeded up, as Bergson famously imagined. Then the entire history of the universe would be unfolded in one stroke, and time could truly be said to be the easiest way nature has found to prevent everything from happening at once.

Suppose, however, that something of the past leaks into the present, and moreover, that past and present differ in kind (in nature), just as the actualized image of past events differs in kind (in nature) from its nebulous counterpart in pure memory. Then there is no reason to think that the present does not contain more than what can be derived from the past: we have made room for genuine creation. This subtle view lies at the core of the idea of duration elaborated in *Creative Evolution*, but it is already at work in the discussion of the relation between present and past in *Matter and Memory*. The same claim can be made from different perspectives.

In a way, Bergson argues that, irrespective of what is actually happening in time, genuine novelty is produced *by the mere fact that things endure*. In other words, relying on common parlance, the passage of time itself is a factor of change. The claim is baffling, and in fact it can only start making sense if we introduce a few qualifications. First, strictly speaking, the present cannot be instantaneous, which means that passage is not adequately described as a dense continuum of instants. Being is a unit of becoming, it must have some breadth. The fact is well confirmed by experimental psychology. However, more profoundly, the thick or 'specious' present should help us grasp what is at the core of Bergson's claim, namely that the past continuously grows upon itself, forcing the present to pass. There is no point denying that the present flows in some sense: the fact that it endures, rather than being given in a flash, makes it less incomprehensible. However, the present does not flow because it is in the nature of time to do so, it does not pass by virtue of conforming to an impersonal and abstract template of 'becoming *in general*' (CE: 324). Bergson suggests that the present passes because it is infused with the past that 'pushes' against it. This is a striking image, but it is no more than that. To spell it out more clearly, we would need to consider what is really involved in the continuity of real change: not so much the mathematical connectedness of continuous magnitudes, but the notion that past and present overlap in an indivisible process, in such a way that the past is genuinely preserved as past within the present – not in actual form, of course, but virtually.

To put it in yet another way, following Bergson's intensive description of the planes of consciousness – the psychic counterpart of the stratification of durations in nature – we may say that the present moment contracts the whole past from its own, non-substitutable perspective. There is an interplay of memory and perception whose joint action makes up the actual present. Otherwise, a present stripped from its concrete content is nothing but the empty form of awareness, or which amounts to the same, the mere form of our activity or 'attention to life'. In fact, to perceive is to remember: the constitutive role of memory in the elaboration of perception is one of the central claims of *Matter and Memory*. As a result of recapitulating the entire past from the perspective of an embodied, situated action, each concrete moment, each slab of experience appears as new in itself by the mere fact that some time has elapsed – that past and memory have accumulated – in between. Thus, each moment is new, and to that extent unpredictable, non-derivable from previous moments. Again, the novelty is not due to particular contents resulting from the combined operation of causal determinations acting from outside, so to speak. Nor is it merely the product of things happening in the present and thus continuously adding new content to reality. Regardless of its content, every moment is new by virtue of the form of time itself. This is the core intuition behind the idea of duration: there is a principle of *conservation* or continuation of the past within the present, amounting to a principle of continuous *variation*. The present draws the whole past with itself and is affected by it. The past is continued in the present, forcing it to change. As a result, no two presents can be identical: even if their objective, material contents were *de facto* indiscernible, they would not carry the same amount of past! This is Bergson's temporal counterpart of the Leibnizian principle of the identity of indiscernibles. The action of time makes for a non-conceptual difference. In the present moment, there is the present *plus* the past, there is perception *plus* memory, contracting the past from a particular perspective.[2] Finally, this explains why the future cannot be simply read off the present, as if nothing was genuinely created in between. Determinism collapses as a general doctrine: it can only have limited scope, and no clear ontological grounding besides the tendency of matter to spread out and behave according to a principle of minimal difference, stemming from duration's state of maximal relaxation or '*détente*'. Bergson argues that deterministic schemes are at best approximations with only local validity: as a matter of fact, these schemes – the laws of nature – only work under certain circumstances, from a particular perspective that enables the scientist to carve out quasi-isolated material systems within the large universe. So much for Laplace.

In what sense does the past 'exist'?

It is important to realize that in this plot, the elusive, transient 'now' is no longer the main character. It is at best a mirage. Bergson, like Whitehead – and even Heidegger, from a different perspective – believed that too much emphasis had been laid by philosophers and psychologists on the present – the 'now' – as the alleged locus of passage. An important step was taken when people like William James derived the philosophical implications of the 'specious present', reflecting on the fact that the lived experience of the present discloses something quite different from what mathematicians picture as a knife-edged instant. However, the more audacious speculative move consisted in acknowledging a *difference in nature* between the present – whatever its shape and breadth – and the past as such. The twofold ontological principle of variation through continuation ultimately turns upon that primitive difference.

In that respect, the cinematographic rendering of becoming as a procession of moments successively coming into actuality misses the point entirely, not only because it suggests a mysterious passage of time – 'at what rate?' one may ask – a motion that turns time into some kind of hyper-process, but more importantly because it takes for granted that present and past are fundamentally homogeneous. The problem with presentism and eternalism is that they both suffer from a deeply engrained attachment to *actualism*: they share the common presupposition that only what is real and actual can be said to exist. By that standard, however, the ontological contents attributed to past and present are fundamentally similar, despite their modal disparity. The past is merely a present that has passed, a demoted present. One is reminded of the image discussed in *Creative Evolution*. Becoming, says Bergson, is made of distinct states or stages, 'set side by side like the beads of a necklace' (CE: 6). We may say that the presentist draws our attention to the beads, which he reviews one at a time, while the eternalist focuses on the solid thread that holds the beads together, in one piece. The difference is immaterial, because both believe in temporal beads.

Or to put it differently, the presentist insists that the past does not exist, but what would the past look like, if it were made to exist? The answer is obvious: it would look exactly like the present that it was. So, if the presentist were to say *what* it is that doesn't exist, the answer would have to be this: what doesn't exist is exactly the same thing which the eternalist says does exists, namely, the moments that were, the present that is now past. Again, the difference between past and present is purely modal.

To better appreciate the shift in perspective introduced by Bergson, we must realize that the question for him is not whether the past exists, but rather whether it is *available*. In other words, the question is how it can be retrieved and whether it can be acted upon, or become a dimension of our present action. The long development concerning unconscious memories in chapter III of *Matter and Memory* (MM: 189 ff) makes a convincing case: there is no serious reason to deny the continued existence of the past – no more reason, at any rate, than to deny the existence of external objects beyond the scope of actual perception. The past exists, in some real sense, below the threshold of consciousness.[3] However, what this really means is that it is not sufficient to describe the past as a present that *was* and is *no more*: the past is not the present *minus* actuality; it is something else altogether. This difference in nature between past and present is the reason why the past does not itself pass but can be said to be conserved in itself as a whole, and to continue itself into the present. It does so not in a deactivated form, as a mere duplicate of past presents lacking only actuality, but in a transformed state which the word 'virtual' is meant to designate.

Thus, reflected in the mind, the passing moment prolongs itself in the present, while growing by involution into the virtual, building momentum as it stretches over multiple levels of consciousness or reality, i.e. coexisting degrees of contraction and dilation in which it is, each time, recapitulated as a whole. To get a feeling of the inherently heterogeneous nature of virtual memories, of all the thresholds of consciousness that must be crossed in order to simply fold back one tiny facet of our past experience onto the present field of awareness,[4] one only needs to focus on the experience of having a word or a name on the tip of one's tongue. Think of all the efforts required to recall its acoustic shape by extracting it from its nebulous state and gradually condensing it into a conscious memory-image (see ME: 192 ff). Considered against the penumbra of virtual becomings that borders each present moment like a halo, the past is not a mere duplicate of the present subsisting in ideal form. It is not a neutralized present. On the contrary, it is the present as usually represented that comes off as the result of a process of subtraction or filtering out of the past. We believe a new present has been substituted for earlier presents that have become obsolete, when that present is in fact the most contracted degree of the past. One might as well say that it does not exist: artificially separated from the past, conceived as an instantaneous cut in the flow of becoming, the present is indeed deprived of all substance.

This idea can be given an even more radical twist by observing that the past, thus construed, is not primarily what comes *after* the present – after the present has passed, that is. As explained in Bergson's quasi-fantastical philosophical essay on the 'memory of the present' (ME: 134 ff), memory is truly simultaneous with perception, to the extent that it does not need to wait to be formed.[5] This bold thesis confirms the description of the 'circles of memory' in *Matter and Memory*: the shortest, most contracted circuit between perception and memory already implies such a contemporaneity between the present and its immediate past (MM: 127). Spelling out the implications of this view would naturally confirm that the dimension of succession is not exactly derivative, but certainly less fundamental for Bergson than what is generally believed. Equally important is the aspect of simultaneity involved at almost every level of time's action.

At this point one might want to object that we are not exactly dealing with the past here, but rather with a particular relation to the past in the form of memory or remembrance. Bergson's answer to this worry is rather straightforward: what other access than memory do we have to the past as such? Granted, it is the proper business of philosophy to investigate the past in itself: not the actual traces of the past, surviving within the present, nor the foregone days projected backwards on the timeline by an act of imagination, and further expressed in language through the grammar of tenses, but the very essence of the past, its sheer pastness. Yet, just as there is no such thing as 'the present' in general, there is no such thing as 'the past' in general. There is no absolute past any more than there is an absolute present, unless it is referred to a particular perspective capable of relating to it. 'Whatever happens at this present moment,' 'whatever happened at the time': such statements hardly make sense. To think of them as short-hand for what the physicist describes as planes of simultaneity slicing across the material universe is artificial at best. That is not our average perspective on time (see note 2). The point is that there is not *one* past, any more than there is a unique duration encompassing all durations. We only have situated access to the past as a whole. In that sense, the relativity of simultaneity is hardly a surprise. There are many pasts and presents, as many as there are present centres of experience in the universe. Accordingly, memory does not primarily *represent* the past; it is rather the past that turns out to be memory – albeit an ontological memory. Bergson's challenge is to extend the conclusions reached on the plane of psychology to the entire material universe: cosmology as 'reversed psychology' (CE: 228).

What is it that passes?

All these remarks point to the same basic intuition: passage is an utterly inadequate metaphor, and it puts undue stress on the successive component of duration. The past does not itself pass; it grows upon itself, allowing for the present to pass and for real change to occur. Seen from this angle, everything falls into place. For the seemingly phenomenological descriptions of 'duration', the peculiar topological nature of pure memory considered as fuzzy or indistinct multiplicity, are merely expressions of a more fundamental fact: the indivisible continuity of change, the continuation of the past in the present, in short their paradoxical *coexistence*, more essential than their depiction as a succession of distinct states, now-present and then-past. In that respect, it is no more illuminating to identify Bergson as the proponent of some variety of the 'growing block theory' of time, than as a presentist or an A-theorist. From his perspective, it doesn't make much difference which side one takes. Once the series of moments (infinitesimal instants or temporal spans) have been laid out on a single ontological plane, we are committed to the view that time is merely the dimension of successive change – a dimension that only needs to be cut out or punctuated in the appropriate way to yield all there is to be said about temporal matters. And that, of course, is the original sin: 'For our duration is not merely one instant replacing another; if it were, there would never be anything but the present – no prolonging of the past into the actual, no evolution, no concrete duration. Duration is the continuous progress of the past which gnaws into the future and which swells as it advances' (CE: 4–5). 'The continuous progress of the past': given what has just been said, such a characterization has nothing to do with the growing block view of becoming, which is just another variety of the cinematographic illusion criticized in *Creative Evolution*, an illusion underlying all conceptions of time and becoming as a global succession of distinct instants or stages.

The cinematographic illusion reconsidered

For what is really at stake in the cinematographic analogy is the way 'passage' acquires a univocal meaning. Bergson's focus is not on the fragmented, discontinuous aspect of the instantaneous film frames, but on the global framing imposed by the uniform motion of the recording and projecting device. The problem is not so much the artificial imitation of movement effected by the juxtaposition of discontinuous stills along the reel, but the presupposition of a single uniform flow embracing a myriad of local flows unfolding across the world according to different rhythms. This global temporal framing is illustrated in the mechanism of the cinematograph by the seemingly automatized run of the film strip at a fixed rate across a beam of light, indifferent to the contents being projected.[6] Its counterpart in scientific practice is nothing but the common usage of coordinate time to locate and order events in a common representational space. Thus, the stroboscopic view of becoming as a series of discontinuous instants (the conception of motion criticized in Bergson's discussion of Zeno's paradoxes) derives from a deeper presupposition regarding the serial nature of the time-dimension. On a different level, the cosmological cinematograph fancied by the metaphysician dormant in every physicist achieves a similar feat, that of bringing together a plurality of heterogeneous durations within a single global mathematical temporal framework. It does so by fitting durations against a homogenous dimension that renders them commensurate, and most importantly enables them to share a univocal understanding of simultaneity – at any rate, in the ideal Newtonian version. As was said before, Bergson argues that this is made possible, at a deeper level, by a cinematographic mechanism *of thought* performing the extraction of a single, simple, impersonal and abstract representation of 'becoming in general' out of the variety of concrete becomings (CE: 324). Such an indefinite, undetermined becoming is not the becoming of anything in particular: its status is that of a universal medium of change. Let us refer to it as 'frame-time'.

Bergson claims this inner cinematograph is unconsciously at work in natural perception even before it is solidified in intellectual schemes. The way he debunks some of its philosophical implications should prevent us from identifying his doctrine of duration as a late modern variety of Heracliteanism. Celebrating 'flow' or 'becoming' in general is not an option here: quite the contrary, such unspecified reference to impersonal time is explicitly criticized as an illusion. In that respect, Bergson would no doubt have been greatly amused to find himself counted in the ranks of the proponents of 'A-time' ... Bergsonism is a philosophy of durations – in the plural. As such, it challenges us to account for their coexistence in a way that does not rely on the abstract concept of simultaneity provided by frame-time, be it in relativized form. That is the real issue behind the cinematographic illusion – not the rehashed complaint lodged against the spatialization of time or the conflation of instants with points on a geometric line. In this regard, the relativistic hyper-cinematograph refracted across innumerable spatiotemporal perspectives, only offers a more sophisticated variety of the same problem.

Simultaneities: another look at the twin paradox

To sum up, everything we have seen so far confirms the intuition that 'passage' and 'flow', far from being primitive and constitutive features of duration, can only be conferred their true meaning once they are referred to the coexistence of different, essentially non-successive aspects of duration. Bergson's main contribution to the understanding of time may have consisted of shifting our attention, through the very concept of duration, to the third, oft-neglected dimension of time identified by Kant as that of coexistence and simultaneity.

Incidentally, it should be clear by now that the philosophical concept of time as the form of real change does not necessarily fall with the criticism of the cinematographic illusion. On the contrary, it gains in precision. Bergson's analysis of the cinematographic mechanism of thought lays bare a crucial problem that could not find its proper formulation in the sole definition of duration. To see what is at stake in the idea of multiple durations unfolding together in time, it seems necessary to inquire into the forms of simultaneity. However, as was suggested earlier, there is no reason why philosophical reflection should confine the meaning of distant simultaneity to the physicist's concept of world-wide instants (i.e. the so-called planes of simultaneity), even in relativized form. Fixing simultaneity relations between space-like separated events by means of appropriate reference frames (i.e. systems of coordinates), implementing this through electromagnetic signalling procedures, is but one way to construe distant simultaneity. The global temporal framing obtained from the use of such devices does not exhaust the meaning of temporal perspectives. As a matter of fact, the relativistic spacetime framework, considered from a purely geometric point of view, already exhibits patterns of simultaneity that are neither global nor strictly non-local. We may refer to them as instances of 'regional' simultaneity. Interestingly enough, they display intrinsic (i.e. frame-independent) characters, in the sense that they can be directly read off from the invariant topological structure underlying the causal order. This agrees with Whitehead's idea that perspectives are not superimposed upon becoming, or imported from outside, but truly embedded in it, as real features of nature.

Spacetime hosts as many regions of simultaneity as there are coexistent processes. Therein lies the significance of the much discussed 'twin paradox', according to which two accelerated observers age differently along their respective journeys.[7] In *Duration and Simultaneity*, Bergson spent much effort distinguishing 'real time' from 'fictitious time', within the very domain of measured time. That is indeed one of the most original insights of the book: what Bergson calls 'real time' is not another name for pure duration, it is a variety of measured time. It is, as he writes, the time measured by a 'real clock' attached to a portion of matter, and to that extent capable of being registered by a living observer. Now as mentioned earlier, some philosophers, following Langevin's suggestion, have identified 'real time' with the physicist's 'proper time', which is at the end of the day a local magnitude measured along time-like curves in spacetime.[8] They were right to the extent that 'real time' is a variety of measured time. 'Duration' in the strict sense cannot be measured, but there is such a thing as time, and time of course can be measured. Bergson's guess is that if measurement is not an entirely spatial affair, then surely duration must be involved at some level. Indeed, it is only by virtue of its relation to lived duration that real time can play the role of a connecting thread, achieving the unification of durations in spite of the discrepancies in measured elapsed times. However, such unification implies more than uncovering relativistic invariants such as proper time or the spacetime interval. It requires the philosophical elaboration of a genuine concept of coexistence. Bergson's appeal to 'real time' takes on its full meaning in this context. Referring to the twins in Langevin's paradox, he writes: 'Not only do the multiple times of relativity theory not disrupt the unity of real time, but they even imply and uphold it. [...] Without this unique lived duration, without this real time common to all mathematical times, what would it mean to say that they are contemporary, that they are contained in the same interval? What could such an affirmation mean?' (DS: 118, modified translation). Thus the 'unity of real time' is confirmed by the 'the simultaneity of flows', and more cogently than any consideration regarding the metrical equality of proper times.

Therein lies the deeper issue behind Bergson's confrontation with relativity. For when one thinks of it, the twins separate only to meet again, and surely it makes sense to say that *while* the traveller was away, cruising in space, his brother got divorced and remarried. Thus, the twins coexist in a strong temporal sense as they go about their business along separate spatiotemporal paths. They may be aging differently, but they never cease to be contemporary as long as they both live. How do we represent this contemporaneity, the temporal togetherness of their independent proper times? Bergson's claim that the 'simultaneity of flows' is in fact more primitive than the 'simultaneity of instants' meets this concern. The question, at this point, is to assess the exact scope or range of the twins' perspectives on the becoming which carries them both. What is the locus of the relational present that they seem to share despite the fact that their respective durations are in some sense incommensurable, and keep dephasing? How far does coexistence extend? How large is a moment of cosmic time, in this particular situation? There is no univocal answer to such questions, because coexistence itself comes in a plurality of modes or regimes – a plurality that is not exhausted by hyperplanes of simultaneity cutting across spacetime at various angles. Spacetime formalism yields definite answers, but it is the task of philosophical analysis to identify the parameters that are genuinely relevant, while addressing the more general issue.

Coexistence and contemporaneity

Coexistence comes in a plurality of modes or regimes which are all embedded and somewhat superimposed within spacetime. In the quotation given above ('they are contained in the same interval ...'), Bergson likens the 'time' elapsed between the moments of separation and reunion – a time which Langevin shows to be measured differently by each twin – to a thick interval of extended present that they both share within what may be called an interval or region of contemporaneity. This thick present is slightly more distended than the one involved in the proximal description of having to wait for the melting of sugar in a glass of water, but the problem is essentially the same.

The situation can be given precise topological meaning in spacetime (Capek 1971: 248 ff), provided we do not forget that the disjoint spacetime paths of the twins remain metrically incommensurable as far as standard simultaneity is concerned. Introducing an inertial frame somewhere in the picture can yield no more than a relative and arbitrary perspective on the overall simultaneity of their unfolding durations: it is frame-time once again, and there is no point trying to deny the relativity of simultaneity defined in that narrow sense, which Bergson rightly associates with the 'simultaneity of instants'. The best one can say is that a continuous one-to-one correspondence between simultaneous events on the twins' respective paths is available in *some* frames (Whitehead 1925: 177). This is already something, because as it happens the very fact that frame-time and global simultaneity relations are available in that way is an absolute (frame-invariant) fact about the situation – a fact that may turn out to be more significant, as far as the 'unity of real time' is concerned, than the much discussed discrepancy between elapsed proper times.

Thus, the Earth twin can 'sweep along' his brother's path at a distance, so to speak, plotting the remote proper time of the cosmic traveller against his own. However, such a device is necessarily relative to the choice of a particular reference frame (the Earth), and essentially disjunctive because the resulting simultaneity relations are neither symmetrical nor transitive and do not yield a continuous and consistent picture of the overall coexistence of the twins. This is a further confirmation that there is nothing absolute, nothing 'real', in the kind of simultaneity achieved by frame-time.

On the other hand, and for all its limitations, the regional simultaneity defined by the topological envelope of becoming bounded by the two events of the twins' separation and reunion nicely illustrates that a notion of simultaneity can be accommodated to make sense of the relational coexistence of two observers sharing a common history. This thick interval defines a sheaf of simultaneity.

Drawing on the light-cone structure of relativistic spacetime, Whitehead devised yet another model for what he called 'contemporary events' or processes. By this he meant those events which are indeterminate as to their time order because they stand in a relation of mutual causal independence (Whitehead 1925: 177). Contemporaneity in this sense implies causal disconnection. This model of coexistence is limited to certain classes of events (spacelike separated, as the physicist describes them) and is not easily generalized to spacetime paths or stretches of duration. Nevertheless, the negative definition of coexistence as disconnection manages to capture an important phenomenological feature of our experience of the extended present: the absence or void which hollows out the time elapsed between the emission of a signal and the reception of a feedback, as when one is somewhat helplessly waiting for an answer to a letter (to a text message in so-called instantaneous communication). It turns out that even the twins, whom we might think of as engaged in ongoing communication, are concerned by this disjunctive form of coexistence. Separated yet contemporaneous, they may well exchange signals continuously (factoring in Doppler effects), but it is as if a certain measure of disconnection in space was intertwined with their connection in time, as if absence was dialectically incorporated within the overall sense of co-presence shared by both. Thus, the twins coexist continuously in Bergson's sense, but on another level they are also contemporaneous with each other in Whitehead's sense despite their communicating in 'real time'.

The importance of perspective and relational time

We could go on reviewing other, more refined models of non-standard simultaneity: there are several others in store, such as the 'active (or interactive) present' based on the Alexandroff interval, which in some sense offers the dual image of Whitehead's contemporaneity (see Balashov 2010). The main lesson is the following: we are not dealing with time, strictly speaking, unless we make room for all its relevant dimensions, including simultaneity in the generalized sense illustrated by the twin paradox. Coexistence, expressed in the grammar of spacetime, already reveals an intricate and multi-layered dialectics of local and global, invariance and perspective, connection and separation. Bergson's aim is of course more ambitious, and in no way limited by the strictures of the spacetime format and its related models. In fact, the notion of proper time underlying the chronogeometry of spacetime is of very little import in his eyes because as a local expression of succession it is utterly devoid of any element of simultaneity. One may argue that, for that very reason, it has nothing specifically temporal about it. At first glance, such a local figure of time, suited to model particular processes unfolding in real time, may appear in tune with Bergsonian duration. Proper time is time measured where the action unfolds, so to speak. It is as if time, finally unfrozen and allowed to flow distributively along every fibre of spacetime, could compensate for the lack of an objective lapse of time between distant events, and the resulting absence of any 'wavefront' of universal becoming. However, succession by itself is barely intelligible along those fibres of time, it cannot be distinguished from a sheer continuum of moments until some sort of perspective comes into play to introduce a notion of 'at the same time,' opening a perspective on time beyond the successive 'here-nows'. In this respect, 'proper time' is at root a four-dimensional concept with no immediate temporal implications. The advocates of the 'block-universe' view of spacetime readily acknowledge this fact: they have no issues describing the block as an agglomeration of timeless fibres and by doing so they prohibit the overall sense of passage associated with global figures of time. Being frame-dependent, such global perspectives are necessarily artificial and relative. The block itself indicates no privileged foliation, it doesn't need one. Hence, the uneasy feeling that something essential has been left out of the picture. It is of course always possible to slide along a particular worldline as if its points constituted a continuous series of 'instants'. No hyperplane is strictly required to do so, and we may think we are thereby animating the static four-dimensional landscape from within, along its fibres, but the point is that in the absence of any natural notion of distant simultaneity at a distance, the succession of instants along a worldline is nothing more than an ordered series. To read earlier-later relations into the series and get a sense of passage, we need some independent justification. A parameter by itself does not provide such a thing. One may use the symbol *t* to enhance the sense of temporal unfolding, but the truth is that such a parameter will do no more than describing (parametrizing) a curve in 4-D space. Time can hardly get more spatialized than that. To argue that since this is carried across 'time-like' intervals in spacetime, one is effectively measuring durations, is merely begging the question.

What is lacking here? Quite simply: a perspective on time. The possibility of 'scanning' local time from a distance. This may involve sweeping along a time-like path by means of a reference frame once that path has been traced out across spacetime (such analysis is by necessity retrospective). Alternatively, it may require real-time signalling procedures and the concrete transmission of information across space, a possibility that is of course available in the twins' scenario, allowing for a continuous (or continuously disrupted) flow of communication between them, even as they are separated. In any case, it is not enough to merely trace or retrace a continuous path by following a parameter as it ranges over an interval. One can flow along with the parameter *t* as much as one likes, mere parametrization will never yield time, much less temporal process. What is needed is an observation deck of some sort, whereby the inner duration of a conscious observer can survey a process unfolding not only in space, but also in time.

To repeat, the main difficulty is that as a local (monotonously increasing) parameter, proper time does not come equipped with any natural notion of simultaneity. Lacking entirely the depth or thickness attached to the idea of real, extensive becoming, it has no real temporal meaning (see Friebe 2012). It only acquires one once we step back and introduce at least a second proper time in the picture, allowing for simultaneity relations to make sense. Jean Piaget's studies on the psychological evolution of the concept of time point in the same direction: temporal categories emerge when the question arises of how to coordinate two (or more) simultaneous flows (see Piaget 2006). Thus, the coexistence of the twins becomes a truly temporal affair as soon as we consider how they can take reciprocal views on their respective durations.[9] As a matter of fact, the actual procedure involved in the physical measuring of any duration already involves just this. What does a clock do, if not plotting a certain motion (or more generally, a sequence of becoming) against another motion taken as reference? The relational account of time defended today by such philosopher-physicists as Julian Barbour or Carlo Rovelli derives from that basic insight.

Conclusion

Time would not 'pass' – we couldn't make sense of its passage – if it were not for the simultaneity of different processes unfolding together. Time by itself does not flow, it can do so only once we recover the extensive character of real becoming, spilling over local happenings to extend across whole regions of spacetime. Non-standard accounts of simultaneity are a valuable cue in that respect; they are instrumental in revealing the dialectic of fibre-time and frame-time, locality and perspective. They force us to account for the joint action of the three fundamental aspects of time – duration, order and coexistence – while throwing light on Bergson's somewhat obscure statements regarding the diffuse, lateral pressure exerted by the whole evolving universe upon its minutest parts: a living consciousness, or a lump of sugar (see CE: 13–14, 369).

Notes

1. 'We are inclined to think of our past as inexistent, and philosophers encourage this natural tendency in us. For them and for us the present alone exists by itself: if something of the past does survive it can only be because of help given it by the present, because of some act of charity on the part of the present, in short – to get away from metaphor – by the intervention of a certain particular function called memory, whose role is presumed to be to preserve certain parts of the past, for which exception is made, by storing them away in a kind of box. – This is a profound mistake!' (CM: 177)
2. Incidentally, this means that the notion of an absolute present, a present that would not be the present of anything in particular, is immediately dubious from a Bergsonian standpoint. Referring to 'the present' without further qualification, as if it were an objective slab of reality, is not only problematic because of Einstein's claims regarding the relative character of simultaneity relations; it is intrinsically misguided because it assumes a homogeneous, impersonal becoming made of successive layers of 'now' stacked up in continuous fashion along an essentially passive dimension of time. When referring to the present, one should always ask: whose present? From which perspective? Keeping this in mind, the claim of a fundamental difference in nature between present and past appears less baffling. 'The past' is not an objective ontological domain composed of all former presents that are now past. In fact these alleged past presents never existed in the first place. Pictured as instantaneous 'planes of simultaneity' cutting across nature as a whole, they are mere abstractions. They are nobody's presents, even if they turn out to be dependent upon the choice of particular reference frames. A reference frame is yet another abstraction, for it amounts to a class of equivalent inertial systems, equipped with appropriate coordinate frames. Only concrete – that is, situated – presents exist, in relation with local conjunctures involving definite durations. The present is 'thick' – as illustrated by the 'specious present' – because it is the present of actual durations, not the mathematical limit of universal becoming. It is not so much a relative concept as a local or regional concept. The same is true of simultaneity.
3. Elsewhere Bergson establishes the continued existence of the past in a more straightforward way by simply emphasizing the continuity of lived duration: 'the preservation of the past in the present is nothing else than the indivisibility of change' (CE: 182); what remains to be established is in what sense the past thus preserved can be said to *exist*.
4. See also MM: 174 and 319.
5. I have commented this idea at length in the French critical edition of 'The Memory of the Present and False Recognition' (Bergson 2012).
6. Interestingly enough, Bergson makes no reference whatsoever to the hand of the operator, at a time where most projectors still involved manual handling. The kind of cinematographic projection he has in mind does not involve any cranking device, nor any possibility of adapting the run of the wheels to the inner rhythm of the filmed action (see During 2015).

7   Paul Langevin's original tale mentions one travelling observer, who on getting back to Earth turns out to have aged less than everyone at home: the difference in the overall elapsed durations depends on the degree to which the travelling observer is accelerated, as well as on the resulting speed. The 'paradox' stems from the apparent symmetry of the observers, who may each consider themselves as rest while the other is moving. According to the principle of relative motion, it may indeed seem that they have equal right to claim that *they* have aged less. Hermann Weyl suggested we choose twins as observers, to further dramatize the issue. The details of this situation are presented in almost every textbook introduction to special or general relativity. Remarkably enough, the strategies adopted to explain away the paradox are almost as diverse as the ways of telling the story. For a genealogical account and philosophical interpretation of the paradox, see During (2014).
8   Contrasting it with the frame-time of the cinematographic illusion, we may refer to this variety of measured time as 'fibre-time' (see During 2015 and 2021).
9   Another illustration of this claim can be found in Bergson's comment on Zeno's paradox of the moving rows (also known as 'the Stadium') in *Matter and Memory* (MM: 252–3), which can be viewed as an anticipation of the relativistic paradoxes. The moral of Zeno's paradox, as far as coexistence is concerned, can be summarized in the following way: there wouldn't be anything *temporal* about either of the lines traced by the same moving body in two distinct reference systems if some consciousness was not capable of surveying the simultaneous unfolding of relative motions and infuse its own sense of duration to the resulting trajectories drawn across space. To achieve this, it is necessary for consciousness to view the whole situation from its own vantage point, at a distance, plotting the various temporal perspectives against each other. That is the fundamental situation to which *Duration and Simultaneity* keeps coming back: an interplay of perspectives attached to moving frames, and the pending question of *where* the philosopher should stand in relation to them. Bergson suggests, and Merleau-Ponty after him: neither *here* nor *there*, but somehow *in between*.

References

Balashov, Y. (2010), *Persistence and Spacetime*, Oxford: Oxford University Press.
Bergson, H. (2012), *Le Souvenir du présent et la fausse reconnaissance*, Paris: Presses universitaires de France.
Capek, M. (1971), *Bergson and Modern Physics: A Reinterpretation and Re-evaluation*, Dordrecht: Reidel.
During, E. (2014), 'Langevin ou le paradoxe introuvable', *Revue de Métaphysique et de Morale*, no. 4: 513–27.
During, E. (2015), 'Notes on the Bergsonian Cinematograph', in F. Albera and M. Tortajada (eds.), *Cine-Dispositives: Essays in Epistemology across Media*, 259–93, Amsterdam: Amsterdam University Press.
During, E. (2021), 'Time as Form: Lessons from the Bergson-Einstein Dispute', in A. Campo and R. Ronchi (eds), *The Quarrel of Time: Bergson vs Einstein*. Berlin: De Gruyter.
Friebe, C. (2012), 'Twins' Paradox and Closed Timelike Curves: The Role of Proper Time and the Presentist View on Spacetime', *Journal for General Philosophy of Science*, vol. 43 (2): 313–26.
Piaget, J. (2006), *The Child's Conception of Time*, London: Routledge.
Whitehead, A.N. (1925), *Science and the Modern World*, New York: The Macmillan Company.

14

Connection and Disconnection of Perception and Memory: Déjà vu, Bayesian and Inverse Bayesian Inference

Yukio-Pegio Gunji
Waseda University

1 Introduction

While Henri Bergson's book *Matter and Memory* (Bergson 1896/2014), which is characterized by the reality of memory and the perception in the luminous point, appears to be far removed from modern brain science and cognitive science, his idea is reasonable with respect to the internal or the first-person perspective and could be consistent with the conscious model proposed by Tononi (2004), Koch (2012), Massimini and Tononi (2013), Dehaene (2014) and Maeno (2010).

However, Bergson's theory could extend beyond modern brain science and artificial intelligence because it focuses on the dynamic interaction between perception and memory, which are different from each other in nature. Such a difference in nature is manifested via the difference between nominalism and conceptualism and is found to be established in connection with a pair of differences. That double axis of differences contributes to the interaction of perception and memory.

Herein, we first show that perception theory in matter and memory is analogous to Gibson's ecological approach to vision (Gibson 1979) and with the passive consciousness hypothesis (Maeno 2010). Second, we show how the present is arbitrarily connected with the past, i.e. the repetition of memory, in the case of déjà vu. The nature of the connection between the present and the past found in déjà vu is also found in our time experience in everyday life. This connection can help us to comprehend the nature of duration as a continuum resulting from an ordered structure of time. Third, we show that the connection of perception and memory can be implemented in the model of inference by introducing a novel process of inverse Bayesian inference (Gunji et al. 2016, 2017, 2018, 2021), where pure recollection and pure perception correspond to Bayesian inference and inverse Bayesian inference, respectively. Finally, we summarize the significance of *matter and memory* in the science of consciousness.

2 Bergson, Gibson and passive consciousness

The system of forming perception is expressed as a network of information propagation in *matter* and *memory*. For Bergson, the spinal reflex is different in degree from perception. He said,

> My body, an object destined to move other objects, is, then, a centre of action; it cannot give birth to a representation.
>
> Bergson 1896/2014, P3L1

The brain modulates information propagation locally in time and space. For Bergson,

> The brain is no more than a kind of central telephone exchange; its office is to allow communication or to delay it. It adds nothing to what it receives.
>
> Bergson 1896/2014, P9L16

A complex network of information propagation can form a local folding structure of information flow in which multiple moments of time are embedded. On this issue, Bergson said,

> However, brief we suppose any perception to be, it always occupies s certain duration, and involves consequently an effort of memory which prolongs one into another a plurality of moments.
>
> Bergson 1896/2014, P11L38

Bergson abandons the distinction and comparison of a real object and its representation and defines a particular universe in which all objects and representations are regarded as images in the network of information propagation. That is a virtual space viewed from the first-person perspective. In this sense, perception is formed in the place of the object being located. Bergson illustrated perception as follows:

> Take, for example, a luminous point P, of which the rays impinge on the different parts, a, b, c, of the retina.
>
> Bergson 1896/2014, P16L18

In this example, perception is not formed in the brain but rather in the place where the point P is located outside of the brain such that the following is obtained:

> The truth is that the point P, the rays which it emits, the retina and the nervous elements affected, form a single whole; that the luminous point P is a part of this whole; and that it is truly in P, and not elsewhere, that the image of P is formed and perceived.
>
> Bergson 1896/2014, P17L14

The image of an object is perceived at the place where the object is located because an object is accompanied by its own surroundings as a whole. This idea is very close to that

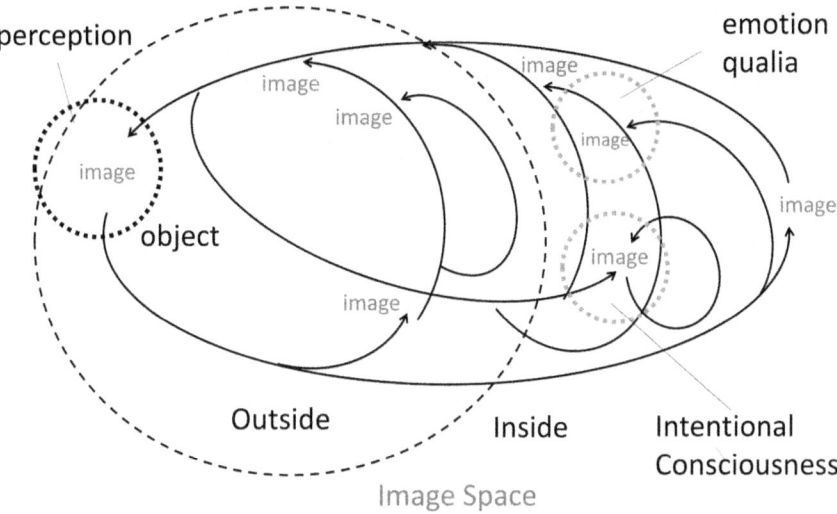

Figure 14.1 Perception of P established in the location of P from the virtual first-person perspective. Emotion and/or *qualia* is formed in the inside. © Yukio- Pegio Gunji.

of Gibson (Gibson 1979). Imagine there is a candle on the floor in front of you. The candle receives the reflections and direct light rays from the room's light and is located on the floor that is extended across the room. Thus, the candle in front of you is seen as a whole view of the room. The candle cannot be separated from its own surroundings. That is, the image of the candle is destined to be perceived at a certain place (Figure 14.1).

This perception is established in the network of information propagation in which any nodes are images and any links are information flows. For Bergson, prolongation and shortcuts related to information propagation can be translated into the asynchronous firing of neurons, which can result in the locally collective behaviour of synchronous firing, according to modern brain science. Although Bergson never claims that intentional consciousness 'knows' about making decisions a posteriori and/ or unconscious initiation, he implicitly claims asynchronous updating is present in the network of information propagation. Bergson also argues that perception results from internal choice. These ideas are consistent with modern brain science and could suggest that passive consciousness is driven by unconscious distributed systems (Koch 2012, Massimini and Tononi 2013, Dehaene 2014, Maeno 2010).

After Libet's experiment on readiness potential, it has been verified that unconscious distributed systems of neural networks could contribute to making a decision before the conscious intention of making a decision presents itself. Brain scientists Koch and Tononi call the unconscious distributed systems consisting of neurons 'zombies', while artificial intelligence researcher Maeno calls the same system 'dwarfs in a brain'. These terms refer to the bottom-up behaviour of neural networks that are independent of intentional consciousness. Zombies (neurons) are divided into populations that are dependent on external stimuli. Each population can result in collective behaviour that reveals synchronous updating and can be regarded as a particular interpretation for a given external stimulus. Finally, the largest region of synchronous firing neurons is chosen to be propagated globally and is consciously presented. This process forms the basis of the global workspace theory proposed by Baars (1989) the global neuronal workspace hypothesis proposed by Dehaene and Naccache (2001).

These studies discriminate unconscious neural activity *a priori* from conscious neural activity *a posteriori*. Brain scientists have tried to untangle the binding problem of how the colour red and a particular aroma are bound together in the image of rose. After the binding of colour and aroma are achieved by unconscious zombies, the conscious region of the brain accesses the result of this binding. Thus, the binding problem is improved.

Bergson also noticed the binding problem:

> In the second, 'the data, of our different senses' are, on the contrary, the very qualities of things, perceived first in the things rather than in us: is it surprising that they come together, since abstraction alone has separated them?
>
> Bergson 1896/2014 P21L30

For Bergson, the idea that the image of an object is perceived at the place where the object is located can prove the binding problem. For example, the image of a candle in front of you is perceived at the exact place in which it is located. That is why binding the object candle with its representation is already completed.

The question remains whether brain scientists' solution for the binding problem equals Bergson's solution for the same problem. While Maeno also considers that the image of an object is perceived at the location of the object, he thinks that such a direct perception is an illusion constructed by the brain. For example, if you scuff the surface of a glass, you feel the smooth-to-the-touch sensation of the glass on the tip of your finger. However, there is only a sensor but not a processor to create the smooth-to-the-touch feeling. Thus, the smooth-to-the-touch feeling is created by labelling presented at location of the finger. In another example, when you watch TV, the voice of a person on the screen is output from the speaker of the TV. You, however, feel that the voice is coming from the mouth of the person on the screen. That is also an illusion created by labelling. Maeno extends this idea and claims that if a feeling is labelled by a particular label, namely, 'at myself', then the feeling is destined to be subjective, i.e., *qualia*. Is this a representationalism criticized by Bergson himself?

> There is not, in fact, an image without real extension which forms itself in consciousness and then projects itself into P.
>
> Bergson 1896/2014, P17L13

Recall the image space proposed by Bergson. Imagine that you are walking and watching a real-time movie that is projected via a head-mount display and recorded by a panoramic camera located by your head. The image of a particular object in front of you is projected in the head-mount display, but you can feel that it is perceived in front of you. That illusion can help you walk in a real space. It is not necessary to check the level of matching between the real location and the virtual location of the object because the real space is isomorphic to the virtual space. I think that this is the essence of the image space proposed by Bergson. In this sense, the distinction between real and virtual space does not make sense. The image space is not a real space in a naïve sense. This universe can be regarded as a virtual space viewed from the first-person perspective.

Bergson criticizes a pure image without a real extension as a representation and checks the match level of the object and its representation. If it is possible to compare an object with its representation and to match something outside with something inside of the brain, this could mean that there is no contradiction and/or friction between inside and outside of the brain. If there is no friction between the inside and outside of the brain, then there is no friction between memory and perception. This possibility gives rise to the destruction of the concept of duration. This is why Bergson criticizes images without a real extension or representation.

The discrimination of an *a priori* process from an *a posteriori* process could establish a division of roles in a brain. On the one hand, unconscious regions in the brain play a role in executing bottom-up choices, while on the other hand, the conscious region plays a role in interpreting and planning a choice a posteriori (Koch 2012, Massimini and Tononi 2013, Dehaene 2014, Maeno 2010). Dehaene has conducted various experiments in which unconsciousness is controlled by using a technique, namely, the priming image technique (Dehaene 2014). If a visual image is presented in a term shorter than several hundred milliseconds, one cannot see the image except for unconsciously. This is called a priming image. Priming images given before a visual image is consciously perceived can control the context of the visual image. If a particular signal follows the priming image of 'push' and another signal follows the priming image of 'do not push', after which you are asked to push a button subsequent to viewing an adequate signal, you do not comprehend why you push a certain button; however, you can adapt to this environment because you can watch the priming image unconsciously. Unconscious priming controls conscious decision-making, but one can feel that one will choose either to push or not push voluntarily.

Although Bergson does not explicitly discriminate *a priori* from *a posteriori*, subjective feeling or emotion can be expressed in his image space. When you see a delicious apple in front of you, the image of an apple is perceived in front of you, and the *qualia* of your deliciousness is perceived inside of your body (Figure 14.1).

Bergson claims that consciousness is just an illusion and that pain and/or emotion is regarded as a local effort to recover the abnormal information propagation:

> The luminous point gives rise to a virtual image that symbolizes, so to speak, the fact that the luminous rays cannot pursue their way. Perception is just a phenomenon of the same kind.
>
> Bergson 1896/2014, P13L38

According to modern brain science, something that cannot be comprehended in consciousness is hidden in unconsciousness. Thus, information in consciousness is just a part of the information reserved in unconsciousness. If this is true, then there is no difference in nature between information in consciousness and that in unconsciousness. This could mean that there is no difference in nature between perception and memory. This idea is far from Bergson's idea.

3 Déjà vu, pure recollection and pure perception

Is recognition a process of checking the match level between an external object perceived now and here with a memory image? To manifest this idea, Bergson discusses the example of déjà vu:

> There are two ways in which it is customary to explain the feeling of 'having seen a thing before'.
>
> Bergson 1896/2014, P47L27

Bergson first assumes that déjà vu results from the incomplete matching of memory images and the view seen here and now. To explain his assumption, he took a model system, i.e. the learning process of walking in an unknown town, to consider the relation between the perception of the present view and memory image.

First, consider walking in an unknown town in which you are just a visitor. At any corner you turn, you cannot predict the view that you will see. Thus, any view in front of you is perceived as something that has not been seen before. Even a view perceived as the unknown is affected by memory images. However, we consider the ideal case, which is pure perception. Bergson says,

> There is, in the first place, if we carry the process to the extreme, an instantaneous recognition, of which the body is capable by itself, without the help of any explicit memory image.
>
> Bergson 1896/2014, P49L23

Second, consider the case in which you are sufficiently used to walking in a town. You know the geography of the town very well, and you have a cognitive map in your head of this town. At any place and at any corner, you know where you are. Does this mean that you are used to walking in the town? If you see just a part of the town, you can imagine the whole town and indicate the location of a part of the town. Thus, you can see the whole town based only on a part of it. In contrast, if you are asked for a representative of the town in the term of a particular context, you can answer and choose a particular part of the town as a representative. Thus, you can see a part of the town as distinguished from the whole. This is a habitat memory that interpolates motion between a part and the whole. The interpolation of a part and the whole can constitute the indefinite whole as a limit of the motion (i.e. a limit of expansion of a part), which is pure memory. Any external stimulus perceived is now completely interpreted in this interpolating motion between a part and the whole. Herein, we consider this interpolating system as containing pure memory and pure recollection.

Third, consider an intermediate case consisting of the end members of pure perception and pure recollection. In this case, objects are perceived through the interpolating system of a part and the whole, and the interpolating system is constituted by adding some fragments of perceived objects. Thus, perception is affected by memory images, and memory images are affected by perception.

Any perception cannot be separated from memory, where the rate of perception and memory varies depending on the condition of perception. Here, we denote this process by perception=memory. If the memory is dominant, then we consider perception=memory to be known; however, if the perception is dominant, then we consider perception=memory to be unknown.

In considering the model of walking in an unknown town, Bergson argues two issues. First, the concept of perception=memory contains both aspects of memory and perception, although they cannot be separated from each other. Second, perception is different from memory in nature; the concepts cannot be compared with each other to check their level of matching. This kind of interpolating system is universal for perception accompanied by memory. The present perceived here and now is limited and constrained under a particular context. When the present becomes the past, this limitation and constraint are degenerated into more indefinite ones. The older the past is, the more open the constraint is. Thus, the memory looks as if it constitutes the order structure. This is the same situation presented in the case of recalling memories while walking in a town because the whole town consists of parts, which demonstrates the inclusion order of parts. The order structure, however, is modified by the interpolation of parts and the whole and by the interpolation of smaller and larger conditions. Thus, it could give rise to an indefinite whole as pure memory and an indefinite instance as the duration of the present.

Under this consideration, recall déjà vu which is not referred to by Bergson himself. Actually, the déjà vu to which Bergson refers is not the same déjà vu that we refer to in the modern situation. In modern cognitive science, déjà vu is explained by the mismatch of a perceived image with the feeling of familiarity. If you perceived a thing before and you feel its familiarity, then you feel that the thing is known to you. Although you have not seen the thing before, when you perceive the thing and you feel familiarity, you experience déjà vu. In the sense of Bergson, a feeling of familiarity can correspond to memory. However, there is no difference in nature between perception and familiarity. Thus, the explanation based on mismatching is inconsistent with Bergson's idea.

How does Bergson explain déjà vu? First, I will explain my private déjà vu experience. One time, I was following an elderly lady who was pulling a cart full of baggage. I saw the wheels of the cart rotating. Long wet lines on the pavement extended from the wheels. At first, I guessed that the rotating wheels were making the trace of the cart since the wheels were wet. Suddenly, the wet lines shifted transversely to the direction of the cart's motion. At this moment, I comprehended that the cart was being pulled by the lady exactly on wet lines that had been made by another cart, and I experienced a strong feeling of déjà vu.

I thought that Bergson might explain déjà vu in the following way: 'At first you believed that the wet lines were made by rotating the wet wheels of the cart being pulled by the elderly lady. It makes sense that "<wet wheels> implies <wet lines on the

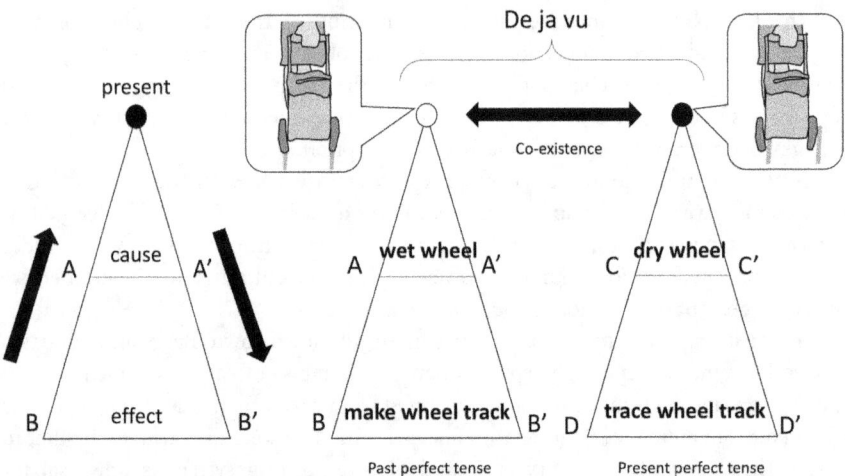

Figure 14.2 An exchange of the past makes of two past histories coexistence, which could create a feeling of déjà vu. © Yukio-Pegio Gunji.

road>", which is recollection=memory image. In the sense of implication, the cause is included in the effect, which results in the order structure.' Bergson might continue to say, 'Simultaneously, <wet wheels> were recognized by the fact that <wet lines on the road> were made. Thus, the inverse implication, "<wet lines on the road> implies <wet wheels>" holds. Both implications and inverse implications can lead to the repetition of the motion: 'Wet wheels cause wet lines on the road. This is nothing but an interpolating system of cause (part) and effect (whole).'

In Figure 14.2, the lines in the triangle horizontal to the bottom line represent cause or effect. If the order that cause implies an effect holds, then the length of the line AA' representing cause is shorter than the length of the line BB' representing effect. The upwards and downwards arrows in Figure 14.2 show that the effect implies the cause, and the cause implies the effect, respectively. These arrows show contraction and relaxation in the interpolating motion in the inverse triangle. The present is represented by a solid circle. In the case of déjà vu mentioned before, the cause AA' corresponds to <wet wheels> and the effect BB' corresponds to <wet lines on the road>. The contraction and relaxation of the triangle implies the repetition of this motion or the cause-effect loop.

Déjà vu shows that the connection of the present and the past (i.e. cause-effect loop) is arbitrary and not necessary (i.e. can be disconnected). Figure 14.2 shows the situation of the appearance of déjà vu in which I saw the baggage cart being pulled by an elderly lady. The cause-effect loop consisting of <wet wheels> and <wet lines on the road> lost the sense of the present when I noticed that the wet lines were shifted transversely to the direction of the cart motion because this cause-effect loop became a 'fallacy past' connecting the present. Thus, the present, which has to be connected with the cause-effect loop of AA' and BB', became absent, which is represented by the blank circle shown in Figure 14.2. However, what happened to the present? The present became connected with the true cause-effect loop such that <dry wheel>, i.e. CC', implies <tracing the wet lines>, i.e. DD'. At the moment that I noticed the change in the cause-effect loop from false to true cause-effect loop, the two cause-effect loops coexisted; i.e. the one was connected with the present, and the other was disconnected with the present.

The cause-effect loop consisting of AA' and BB' lost the support of the present, which was left suspended in an unsupported manner. Then, a cause-effect loop consisting of CC' and DD' newly appeared and acquired the support of the present. When the cause-effect loop (i.e. history) is connected with the present, the history has the present perfect tense. If the cause-effect loop is disconnected from the present, then it is regarded as having been experienced before; this is the past perfect tense. If the false perception (i.e. the cause-effect loop consisting of AA' and BB') disappears quickly, then the feeling of déjà vu never appears. The coexistence of the present perfect tense and the past perfect tense can give rise to the feeling of déjà vu.

It is easy to see that the past is sometimes connected with the present but also sometimes disconnected from the present when considering déjà vu. The relation between the present and past is, therefore, arbitrary. If various inverse triangles corresponding to different cause-effect loops are collected and synthesized, one can obtain not an inverse triangle but rather an inverse cone (Figure 14.3). This inverse cone contains various cause-effect loops.

The connection between past and present is dependent on the weight of cause-effect loops. Sometimes the flow from the past (cause-effect loop) can rush towards the

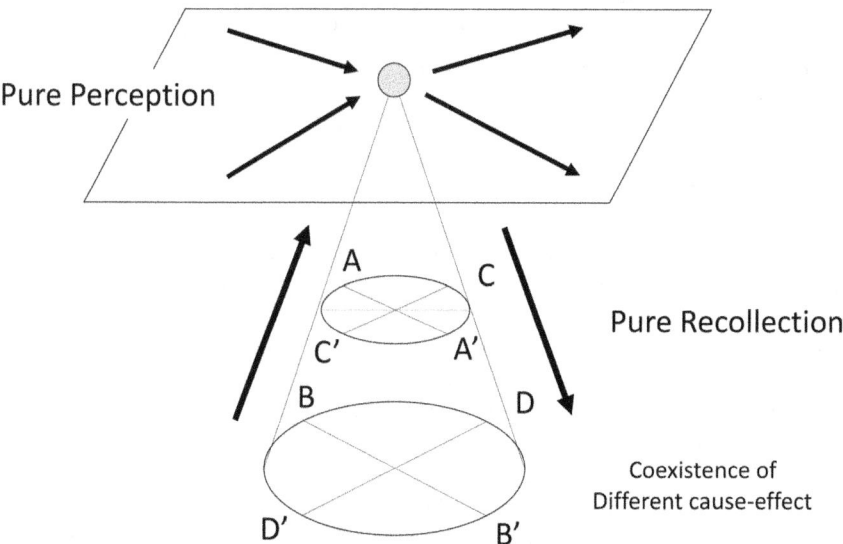

Figure 14.3 Inverse cone generated by the synthesis of multiple triangles and planes. © Yukio-Pegio Gunji.

present, and sometimes the connection between past and present is cut off, dependent on the weight of cause-effect loops. Thus, perception=memory cannot be divided into pure perception, in which the present is collected from a plane, and pure recollection.

4 Bayesian and inverse Bayesian inference

In this section, the structure of perception=memory is manifested by referring to the dynamic relation between conceptualism and nominalism in *Matter and Memory*. The structure is implemented herein by the inference model based on Bayesian and inverse Bayesian inference (Gunji et al. 2016, 2017, 2018, 2021).

First, we will summarize the idea of Bayesian and inverse Bayesian inference. The essence of Bayesian inference is removing redundant possibilities in searching for the optimal solution. If one were to search for the optimal solution over a whole range of possibilities, it would take a huge amount of time. Since Bayesian inference removes unexperienced possibilities, the candidates of the optimal solution are reduced only to experienced possibilities, which can help one achieve more efficient searching. This implies that the space of searching for an optimal solution is degenerated into a set of steady states or habitualized states. Therefore, Bayesian inference cannot handle unexpected events. Inverse Bayesian inference is implemented to compensate for this weak point of Bayesian inference. Although solo Bayesian inference never allows the change of hypotheses by which inference is realized, inverse Bayesian inference can modify hypotheses to handle unexpected events. Here, we show that a pair of Bayesian and inverse Bayesian inferences results from Bergson's perception=memory.

Bergson argued regarding the conflict between nominalism and conceptualism, which could be superimposed in the structure of parts and the whole in pure recollection. Afterwards, the axis of the general and individual orthogonal to the axis of parts and whole is discovered. This process refers to the connection of pure recollection and pure perception.

According to nominalism, everything is expressed as an extent, while for conceptualism, everything is expressed as intent. These definitions sound as if the concepts are different in their nature and are thus alternative. Bergson, however, argues that the general idea explained by nominalism needs conceptualization, intent and the help of conceptualism. Thus, one can see a contradictory oscillation between nominalism and conceptualism. Soon after finding an oscillation, Bergson says that oscillation is established only if there is a common base of nominalism and conceptualism. This refers to a particular specific framework in which there is a common base and to an axis of multiplication in which various relations of nominalism and conceptualism are indicated. For Bergson, this axis is conceptualized as similarity that is different from the individual and the general. Bergson says that via this similarity (i.e. via the axis of multiplication), one can escape from a contradictory oscillation.

The contradictory oscillation of nominalism and conceptualism and of extent and intent is compared to the interpolation of parts and the whole. Since the interpolation of parts and the whole could destroy the order, this interpolation can result in a dynamic connection to the pure perception so as not to be destroyed. By introducing similarity (i.e. the axis of multiplication), one can escape the simple interpolation of parts and the whole. In other words, the intervening of pure perception in pure recollection could correspond to introducing the axis of multiplication. This implies that the connections of the present and past are different from each other in nature. It also enables the motion not to converge to a contradiction.

Under this consideration, the inverse cone shown in Figure 14.3 is implemented in the inference model. Recently, many researchers in cognitive science have focused on Bayesian inference (Bayesian 1763/1958). In particular, Friston (2006; 2009a, b; 2010) has generalized the idea of Bayesian inference in the form of the free energy minimization principle in which the Kullback–Leibler divergence between the *a priori* and the *a posteriori* probability is minimized, which leads to Bayesian inference. One can use Bayesian inference to make a decision, where the subjective probability distribution changes depending on experience. Imagine the probability of succeeding in making your pet cat say 'meow'. Without any information, you might assume that the probability is 0.5. However, if you place dry food in front of your cat, the probability becomes 0.8. Showing the cat dry food represents a particular condition. For normal probability theory, the probability under a specific condition is discriminated from the probability without a condition. In contrast, for Bayesian theory, the probability under a specific condition is regarded as the *a posteriori* probability of the probability without conditions. In this sense, the probability of making your cat say 'meow' changes from 0.5 *a priori* to 0.8 *a posteriori*. By introducing the distinction between *a priori* and *a posteriori* probability, Bayesian theory identifies the probability without conditions as being different from the probability under certain conditions.

Imagine a case in which there are red balls and white balls, with a total of 10 balls in a bag. You take one ball out of the bag, confirm whether it is white or red, and then return it to the bag. You have two hypotheses for this bag. The first hypothesis consists of 3 red balls and 7 white balls, and the second hypothesis consists of 7 red balls and 3 white balls. If you have a series of balls taken from the bag such as red, red, red, etc., which represents the repetition of red balls, then you might think the second hypothesis is reasonable. For Bayesian inference, the probability of the first and the second hypothesis is initially assumed to be even. A knowledge of a series of balls corresponds to a condition. Thus, the probability of the first and the second model occurring without a condition is replaced by the probability under a condition of this knowledge; then, the probability of the second hypothesis being adequately chosen increases step by step.

Bayesian inference can be regarded as an implementation of the above-mentioned global workspace theory. The global workspace theory is an optimization process in which the largest synchronized neural region is chosen for a given external stimulus. Since neural regions and external stimuli correspond to the hypothesis and data, respectively, the choice of the optimized region can correspond to the calculation of the probability of the hypothesis under a condition of given data. For Bayesian inference, probability is perpetually replaced by conditional probability. Optimization is accelerated by modifying the probability distribution of the hypothesis, which is based on the replacement of probability with conditional probability. In other words, optimization is verified by the free energy minimization principle (Friston 2006; 2009a, b).

While Arecchi (2007; 2011) agrees with Bayesian inference in brain science, he claims that we have to manifest hidden judgement processes in the brain. He argues that after calculating probability *a posteriori*, the brain is destined to modify the *a priori* probability to be consistent with the a posteriori probability before replacing the a priori probability with the a posteriori probability. This is called inverse Bayesian inference. However, it is unclear how to implement his idea in a dynamical inference model. Regardless, I think that both Bayesian and inverse Bayesian inference can be implemented in Bergson's inverse cone. While the two concepts are symmetrical to each other to some extent, the symmetrical structure is destroyed when connected to pure perception (Gunji et al. 2016, 2017, 2018, 2021).

Figure 14.4 shows the Bayesian and inverse Bayesian inference implemented in the inverse cone. As mentioned before, the cone represents the order of the inclusion relation. The smaller the part included in the larger one is, the higher the one in the inverse cone is. In this sense, the conditional probability included in the probability is located above the probability located in the inverse cone. The symbol $P(B)$ represents the probability of an event B, and $P(B|A)$ represents conditional probability B under the condition in which an event A occurs. Recall the inference made about the bag containing the white and red balls. Event D corresponds to, for example, a specific datum of taking a red ball, and event H corresponds to, for example, choosing the first hypothesis (called hypothesis H). Here, $P(H|D)$ implies the probability of hypothesis H occurring under the condition of observing datum D. The bottom plane of the cone represents the probability of arbitrary events that can occur. Thus, it is the probability of universal set (U) and is trivially 1.0. The probability of arbitrary events contains every event that has not occurred. This can correspond to pure memory.

As mentioned earlier, the inverse cone refers to the interpolation of parts and the whole, i.e., of contraction and relaxation. In Figure 14.4, contraction, which is represented by the upward arrow, is defined by replacing $P(H)$ with $P(H|D)$. Relaxation, which is represented by the downward arrow, is defined by replacing the conditional probability $P(D|H)$ with the probability of $P(D)$. The contraction exactly corresponds to Bayesian inference, since the probability of the hypothesis is replaced by that of the hypothesis under a specific condition. This implies the degeneration of the experienced possibility. Relaxation corresponds to the inverse Bayesian inference. While the inverse Bayesian inference is defined symmetrically to the Bayesian inference, the likelihood of the hypothesis being nothing but the definition of the hypothesis $P(D|H)$ is replaced by $P(D)$, which is the probability of a specific datum. Inverse Bayesian inference cannot be closed in the inverse cone because the probability of D that is an empirical datum is unknown for hypothesis H. On the one hand, for the Bayesian inference from $P(H)$ to $P(H|D)$, all hat one has to do is forget the unexperienced conditions. On the other hand, for the inverse Bayesian inference from $P(D|H)$ to $P(D)$, one has to remake hypotheses as the engine of the inference. Thus, the inverse Bayesian inference cannot be closed in the inverse cone.

With the inverse cone, the interpolation implements only the substitution of the probability, with no process of choosing an event. That is why only the inverse cone does not contain the inverse Bayesian inference. The calculation of probability can be implanted in the inverted cone, i.e. pure recollection. The choice of event and/or the

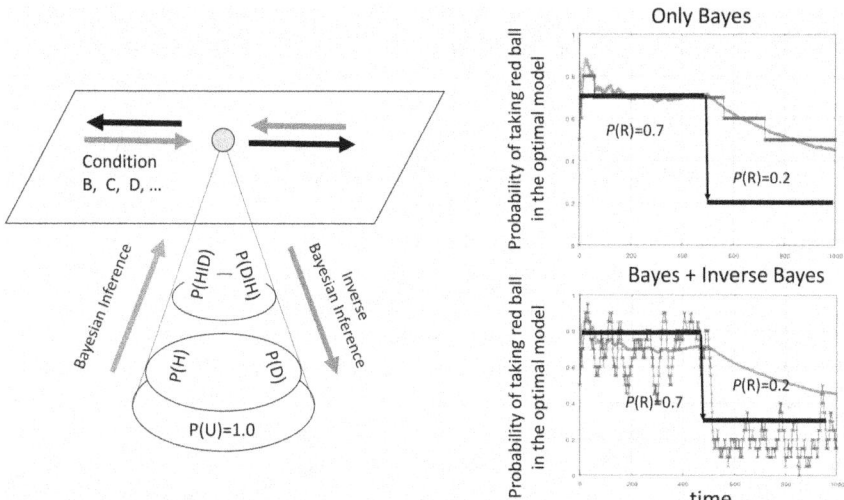

Figure 14.4 Left: inverse cone implemented by Bayesian and inverse Bayesian inference. Right top: the probability of taking a red ball from the optimum hypothesis estimated only by Bayesian inference potted against the time step. The probability of taking a red ball is changed from 0.7 to 0.2 when plotted against time. Right bottom: the probability of taking a red ball of the optimum hypothesis estimated by Bayesian and inverse Bayesian inference when plotted against time. © Yukio-Pegio Gunji.

calculation of the probability of datum is searched for in the plane connected with the cone. That is pure perception. By using the help of pure perception, one can choose the datum and calculate the probability of the datum (Figure 14.4). Thus, the connection between the plane and the cone is implemented by the interaction between events and probability. In the downward interpolation, inverse Bayesian inference affects the plane, and the probability of an event (i.e. condition) can be expanded. This results in the replacement of the hypothesis itself. Thus, the inverse cone affects the plane that is the choice of condition, and the plane affects the cone that is the replacement of the hypothesis itself. The perception induced by inverse Bayesian inference affects Bayesian inference and vice versa.

Figure 14.4 (right) shows how inverse Bayesian inference can contribute to estimating the time series of data. Recall the bag that contains red and white balls. There are many hypotheses to match the real data, where the probability of taking a red ball suddenly changes from 0.7 to 0.2 when plotted against time. This is represented by the stepwise black curve shown in Figure 14.4 (right). The red curve represents the accumulated probability of taking a red ball. At each time step, one takes one ball from the bag and then calculates the probability of taking a red ball. The blue curve represents the probability of taking a red ball based on the optimized hypothesis. The blue curve in the right above diagram shows the probability estimated only by Bayesian inference. That inference is fit to the curve of accumulated probability of taking a red ball but cannot be fit to the actual sudden change of the probability. The blue curve on the right below the diagram shows the probability estimated by both Bayesian and inverse Bayesian inference. By using this inference system, one can immediately adapt to the actual sudden change of the probability, where it can generate autonomous perturbation.

The relationship of Bayesian and inverse Bayesian inference can reveal Bergson's inverse cone. While the inclusion order of conditions is assumed, the probabilities under a condition and those probabilities without a condition constitute the interpolating structure that reveals anticipation and interpretation. The base plane of the cone refers to pure memory. Although Bayesian inference and inverse Bayesian inference appear symmetrical, they are asymmetrical. Only inverse Bayesian inference can affect the plane of events, which is pure perception; pure perception could affect the interpolation of the probabilities by replacing the hypothesis. The operating probability, which is expressed as a real number, is implemented in the inverse cone, while the operating event, which is expressed as a natural number, is implemented in the plane. Thus, the difference in nature between perception and memory can be implemented in the relationship between Bayesian and inverse Bayesian inference, and the dynamic interaction between them is also well implemented. This basic structure is consistent with Bergson's basic idea, which looks as if it is based on the dualism of matter and spirit and/or of perception and memory but rather extends beyond such dualism because of the dynamic interaction between the concepts. The author implements this idea not only for human decision-making (Gunji et al. 2016; 2017) but also for animal decision-making (Gunji et al. 2018, 2021). There can be various ways to implement inverse Bayesian inference. The author has previously implemented inverse Bayesian inference by using rough set approximation (Gunji et al. 2016). This process shows that the logical structure in a cognitive system can be expressed as an orthomodular lattice that is nothing but quantum logic. It has also been shown that the pairing of information contraction inside the focal context with information expansion outside the context entails quantum logic (Gunji et al. 2020, Gunji and Haruna 2022, Gunji and Nakamura 2022a, b). Since the pairing of contraction and expansion is analogous to the pairing of Bayesian and inverse Bayesian inference, such an information structure might generalize macroscopic decision-making.

5 Conclusion

To overcome the dualism of matter and spirits, or of objects and representation, Bergson proposes the intermediate concept of an image. However, what is an image? Herein, we have interpreted the image space as a virtual space viewed from the first-person perspective. Since there is no distinction made between an object and its representation, no check the matching level of the object and its representation is needed. Since the virtual space is isomorphic to the real space, one can see the virtual space as the first-person perspective. When Bergson's image space is interpreted as the first-person virtual space, it is easy to see that the image of the object is perceived at the location where the object exists. Pain, emotion and *qualia* are perceived at a particular place inside of the body in the image space. By defining both the inside and the outside, subjective feeing can be explained dependent on the location of perception.

Although an image space looks similar to a kind of holistic view, i.e. a universe in which a subject cannot be separated from its outside, an image space that contributes perception is connected with memory; this is neither a dualism nor a holism. Rather, it focuses on the dynamic interaction between perception and memory.

In matter and memory, memory is first comprehended as habitat memory and then regarded as the recollection of memory. On the one hand, the motion of recollection can constitute the order structure, while, on the other hand, a particular structure that is contradictory to the order structure can lead to pure memory. We manifest this kind of structure as the interpolating system between parts and the whole. Since the implication is also expressed in the inclusion order, the habitat causal relation is also interpreted as the interpolating system between cause and effect, where cause implies effect and vice versa. This can be expressed as an inverse cone.

Via consideration of the déjà vu experience, we explain herein the arbitrary connection of the present and the past (i.e. the inverse cone or the interpolating system of parts and the whole). The past is the repetition of the cause-effect loop. If this cause-effect loop is consistent with the present, then the past is connected to the present. Otherwise, the past is disconnected from the present. If the past that is disconnected from the present is reserved at a particular duration and another past is simultaneously connected with the present, then one can feel a déjà vu experience due to the coexistence of these two pasts (two cause-effect loops). This mechanism can be adapted to the general recollection of memory that is dynamically connected with perception.

Finally, we manifest the relation between perception and memory by implementing the relation of Bayesian and inverse Bayesian inference. Bayesian inference can be implemented as the interpolating system of probability, which is expressed as replacing the probability with the conditional probability. Inverse Bayesian inference is not closed in the interpolating system and is connected to calculating the probability of events in a plane. Via the implementation of inverse Bayesian inference, the dynamic connection of perception and memory is comprehended as the inverse cone that is connected to the plane.

Bergson's theory of perception and memory is consistent with modern brain science and is practical for constructing an artificial perception system not from the third-person perspective but rather from the first-person perspective. Bergson's *Matter and Memory* is very useful for researchers in the fields of artificial intelligence, theoretical biology and cognitive science.

This work was supported by JSPS-JPJS00120351748.

References

Arecchi F.T. (2007), 'Physics of Cognition: Complexity and Creativity', *The European Physical Journal Special Topics* 146: 205.

Arecchi F.T. (2011), 'Phenomenology of Consciousness: From Apprehension to Judgment', *Nonlinear Dynamics, Psychology and Life Sciences*, 15: 359–75.

Baars, B.J. (1989), *A Cognitive Theory of Consciousness*, Cambridge, MA: Cambridge University Press.

Bayesian, T. (1763/1958), 'An Essay Toward Solving a Problem in the Doctrine of Chances', *Philosophical Transactions of the Royal Society of London 53*: 370–418 (second publication is at *Biometrika*, 45: 296–315).

Bergson, H. (1896/2014), *Matter and Memory*, trans. N.M. Paul and W.S. Palmer, Poole: Solis Press.

Dehaene, S. and L. Naccache (2001), 'Towards a Cognitive Neuroscience of Consciousness: Basic Evidence and a Workspace Framework', *Cognition*, 79: 1–37.

Dehaene, S. (2014), *Consciousness and the Brain*, New York: Brockman Inc.

Friston, K.J., J. Kilner and L. Harrison (2006), 'A Free Energy Principle for the Brain', *Journal of Physiology*, 1000: 70–87.

Friston, K.J. and S. Kiebel (2009a), 'Cortical Circuits for Perceptual Inference', *Neural Networks*, 22: 1093–104.

Friston, K.J. and S. Kiebel (2009b), 'Predictive Coding Under the Free-energy Principle', *Philosophical Transactions of the Royal Society of London, B, Biological Sciences*, 364: 1211–21.

Friston, K.J. (2010), 'The Free-energy Principle: A Unified Brain Theory? *Nature Reviews Neuroscience* 11: 127–38

Gibson, J.J. (1979), *The Ecological Approach to Visual Perception*, Boston: Houghton.

Gunji, Y.P., K. Sonoda and V. Basios (2016), 'Quantum Cognition Based on an Ambiguous Representation Derived from a Rough Set Approximation', *Biosystems*, 141: 55–66.

Gunji Y.P., S. Shinohara, T. Haruna and V. Basios (2017), 'Inverse Bayesianian Inference as a Key of Consciousness Featuring a Macroscopic Quantum Logic Structure', *BioSystems* 152: 44–63.

Gunji Y.P., H. Murakami, T. Tomaru and V. Vasios (2018), 'Inverse Bayesian Inference in Swarming behavior of Soldier Crabs, *Philosophical Transaction of the Royal Society* A 376: 2017.0370.

Gunji, Y.P., K. Nakamura, M. Minoura and A. Adamatzky, (2020), 'Three Types of Logical Structure Resulting from the Trilemma of Free Will, Determinism and Locality, *BioSystems* 195: 104–51. DOI information: 10.1016/j.biosystems.2020.104151

Gunji, Y.P., T. Kawai, H. Murakami, T. Tomaru, M. Minoura AND S. Shinohara (2021), 'Lévy Walk in Swarm Models Based on Bayesian and Inverse Bayesian Inference', *Computational and Structural Biotechnology Journal* 19: 247–60.

Gunji, Y.P. and T. Haruna (2022), 'Concept Formation and Quantum-like Probability from Nonlocality in Cognition', *Cognitive Psychology*, DOI.org/10.1007/s12559-022-09995-1

Gunji. Y.P. and K. Nakamura (2022a), 'Psychological Origin of Quantum Logic: An Orthomodular Lattice Derived from Natural-Born Intelligence without Hilbert Space', *BioSystems* 215–16, 104649.

Gunji, Y.P. and K. Nakamura (2022b), 'Kakiwari: The device Summoning Creativity in Art and Cognition', in A. Adamatzky, *Unconventional Computing, Philosophies and Art*, Singapore: World Scientific.

Koch, C, (2012), *Consciousness*, Cambridge, MA: MIT Press.

Libet, B., E.W. Wright Jr AND C.A. Gleason (1982), 'Readiness-potentials Preceding Unrestricted "Spontaneous" vs Pre-planned Voluntary Acts, *Electroencephalography and Clinical Neurophysiology*, 54: 322–35.

Libet, B., E.W. Wright Jr, C.A. Gleason and D.K. Pearl (1983), 'Time of Conscious Intention to Act in Relation to Onset of Cerebral Activity (Readiness-potential), the Unconscious Initiation of a Freely Voluntary Action, *Brain* 106: 623–42.

Libet, B. (2004), *Mind Time: The Temporal Factor in Consciousness*, Cambridge, MA: Harvard University Press.

Maeno, T. (2010), *Why Brain Makes 'Mind'?* Chikuma Press (in Japanese).

Massimini, M. and G. Tononi (2013), *Nulla di piu grande*, Milan: Baldini and Castoldi.

Tononi, G. (2004), 'An Information Integration Theory of Consciousness, BMC *Neuroscience*, 5 (42) doi:10.1186/1471-2202-5-4

The Extensionalist View and Bergson's Notion of Contraction

Ryusuke Okajima
Niigata University

Introduction[1]

Two chapters included in this collection, B. Dainton's 'Neutral Monism, Temporal Experience and Time' and Y. Hirai's 'What is the *Thickness* of the Present?' position Bergson as an Extensionalist with respect to recent debates regarding temporal consciousness. However, it seems that they demonstrate different views on the Bergsonian notion of contraction: the former regards this notion as a remnant of Retentionalism, but the latter assigns to it an Extensional account. This chapter offers an Extensional interpretation of contraction that differs from Hirai's and is inspired by Dainton's overlap theory.

In Section 1, I briefly examine some premises for this notion. In Section 2, after identifying three functions of contraction (synthesis, condensation and reduction), I clarify the interrelationships among them. In Section 3, I present an interpretive model to give a consistent account of all of these functions.

1 Premises of the notion of contraction

Let us begin by introducing some metaphysical and epistemological premises. First, the notion of contraction is based on (at least) two metaphysical assumptions. (1) According to Bergsonian *panpsychism*, matter demonstrates, in itself, a certain phenomenal character and experiential duration. (2) Further, his *direct realism* affirms that we directly experience (i.e. without intervening representation) matter itself. Both these assumptions must be explicated, but here, I move forward without entering further into them.

From an epistemological point of view, it is important to realize that contraction concerns not the *experience of succession* – which is usually investigated by theories of temporal consciousness – but rather the constitution of a minimum conscious moment that forms the basis of such experience. Let us clarify this difference with reference to auditory experience. For two sounds to be heard as successive events, they must be separated by at least 30–40 milliseconds (the *order threshold*). Without this separation, we cannot tell which occurs first. Where the temporal span of two successive auditory stimuli is between 2–3 milliseconds and 30–40 milliseconds, they are heard a simultaneous but distinct. If the span drops below 2–3 milliseconds (the *coincidence threshold*), they become indistinguishable (i.e. heard as one, not two).[2] According to the classical experiment conducted by S. Exner, to which Bergson (MM: 272[231]) and James both refer, the doubleness of two successive snaps is heard distinctly when their interval is 0.00205 seconds, but they become 'a single-seeming sound' when the interval is 0.00198 (James 1950: 613–14).[3] Given this fact of experience, theories of temporal consciousness generally treat the experience such that the timescale is greater than the order threshold. In doing so, they deal with the structure of the experience of succession by accepting the *succession of experience* as a given. However, this situation is different for Bergson. Because most audible sounds are (from the physical perspective) vibrations whose cycle is less than 2 milliseconds, to maintain that we experience them directly, we are required to answer the question: how do these vibrations, which in themselves occur successively, become one single-seeming sound? The notion of contraction was developed to respond to this very question.

2 Functions of contraction: synthesis, condensation, reduction

Keeping these premises in mind, let us discriminate between three functions of contraction. While we referred to the auditory experiences above, Bergson provides an explanation of this notion by appealing to the visual experience of colour. Therefore, I will also use the visual experience as an example (however, the interpretation presented below also holds, I think, for auditory experiences as well). Bergson's discussion for our purposes begins as follows.

> May we not conceive, for instance, that (2) the irreducibility of two perceived colours is due mainly to the narrow duration (1) into which are contracted the billions of vibrations which they execute in one of our moments? If we could stretch out this duration, that is to say, live it at a slower rhythm, should we not, as the rhythm slowed down, see these colours pale and lengthen into successive impressions, still coloured, no doubt, but nearer and nearer to coincidence its pure vibrations?
>
> MM: 268–9[227–28]; number added

Therefore, the operation of contraction (1) constitutes one of our moments from billions of material vibrations and (2) forms our sensible (or perceptual) quality from material sensible quality. Let us call these two functions, respectively, *synthesis* and *condensation*. Although Hirai also provides a clear discussion of these, here, I emphasize the additional and decisive function appearing here in the operation of contraction. Below is a crucial passage for our interpretation.

> In the space of a second, red light ... accomplishes 400 trillions of successive vibrations. If we would form some idea of this number, we should have to separate the vibrations sufficiently to allow our consciousness to count them, or at least to record explicitly their succession; and we should then have to enquire how many days or months or years this succession would occupy. Now, the smallest interval of empty time which we can detect equals, according to Exner, 1/500 of a second. ... Let us imagine, in a word, a consciousness which should watch the succession of 400 trillions of vibrations, each instantaneous, and each separated from the next only by the 1/500 of a second necessary to distinguish them. A very simple calculation shows that more than 25,000 years would elapse before the conclusion of the operation. (3) Thus, the sensation of red light, experienced by us in the course of a second, corresponds in itself to a succession of phenomena which, separately distinguished in our duration with the greatest possible economy of time, would occupy more than 250 centuries of our history.
>
> MM: 272-3[230-31]; number added

Thus, contraction (3) shortens the enormous duration of overall material vibrations. Let us call this function *reduction*.

To make our argument explicit, let us clarify the three functions of contraction (synthesis, condensation, and reduction) using an example based on the above citation.[4]

Suppose I am gazing now at a red patch located at point **P**. While observing it, I have various experiences (such as visual perceptions of other points in space, inner bodily sensations, recollections of past events or abstract streams of thoughts), and I ignore all of them to consider only experience **E**, namely, a visual perception of red light that persists only for a second. Suppose, following the assumption cited above, that the single moment **M**, a minimum conscious constituent of **E**, exhibits temporal extension equal to 1/500 of a second. According to Bergson, we can obtain the experience **E** by contracting 400 trillion successive vibrations of red lights. Based on his panpsychism assumption, we can call **v** a single successive vibration of these lights, and **m** a moment, the temporal extension of which is 1/400 of a picosecond where **v** itself feels only its single vibration, and **e** is the material experience, made up of 400 trillions of moments **m**.[5] In this case,

1. Synthesis is the process whereby each **M** is constituted from 800 billion (400 trillion/500) **v**s.
2. Condensation is the process whereby the quality of perception of red (that each **M** takes on) is formed from the material qualities of 800 billion **v**s.
3. Reduction is a process whereby 400 trillion **v**s are experienced for enormously shorter during in **E** than in **e**.

3 Interpretation of the notion of contraction: Extensional contraction models

We have only been concerned with plural vibrations, but we should consider two conditions to understand how the contraction operates on *each* vibration. Because (a) we adopt *panpsychism* and *direct realism*, each vibration must be able to be experienced without its own sensible quality or duration being altered. However, (b) according to *Exner's experiment*, we cannot experience any temporal extension of less than 1/500 of a second. At first glance, these conditions seem to be incompatible with each other; because the period of the relevant vibration is below 1/500 of a second, if we affirm (b), it appears that the duration of the vibration must be modified; conversely, if (a) is accepted, we experience a temporal extension of less than 1/500 of a second. However, we can satisfy these conditions by providing the following interpretation of the act of contraction: *when it is contracted, each vibration is experienced for an objectively longer period than before but subjectively by the same duration*. To use the example above, this amounts to saying that each **v**, being contracted, is experienced for 1/500 second, but no change is found in the subjective time-length (i.e. the duration itself) before or after the contraction.

This final phrase requires some clarification, as the expression time-*length* seems an inappropriate tool to use to understand the Bergsonian duration, which is, by definition, immeasurable. Note, however, that Bergson does not deny a correspondence between a physical time and a mental duration. Indeed, in the passage referred to above, he relates a unit of time (1/500 second) to a duration experienced during that unit of time. Here, what is important is the distinction between the physical unit of time and the durational unit of time, and the duration as time-length contains no contradiction if this term is construed in the latter sense.[6]

The three operations of contraction described above are, in fact, the results of the act of contraction for each vibration. In what follows, I explain this point in contrast with Hirai's interpretation of the notion of contraction.

(1) First, consider how the function of *synthesis* results. With respect to the above example, Figure 15.1 represents the succession of material moments. In this figure, **v1, v2, v3, ... vα** represents a single vibration of red light. Similarly, **t1, t2, t3, ... tα** represents the beginning of each vibration, and **m1, m2, m3, ... mα** represents a moment where each vibration only experiences itself (herein, α = 800 billion. Note that there are countless vibrations after **vα** in reality).

Figure 15.2 expresses a state wherein all of the vibrations depicted in Figure 15.1 are contracted. We can see that two vibrations are experienced simultaneously from **t2** to **t3**, and three vibrations are experienced from **t3** to **t4**. Similarly, after **t4**, the number of vibrations experienced simultaneously continues to increase (the dotted rectangles express the fact that there are much more vibrations in reality), and at **tα**, it naturally reaches 800 billion. However, from **tα** on, this number remains the same, even if the contraction continues because, after **tα+1**, the vibration that began to be experienced 1/500 seconds ago drops from our consciousness at the same moment that a new vibration begins to be experienced (when **vα+1** arises, **v1** passes away, etc.). After **tα**, each vibration is always experienced partially, simultaneously with other α-1 vibrations:

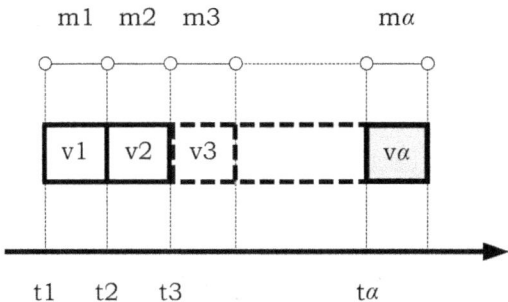

Figure 15.1 The succession of material moments. © Ryusuke Okajima.

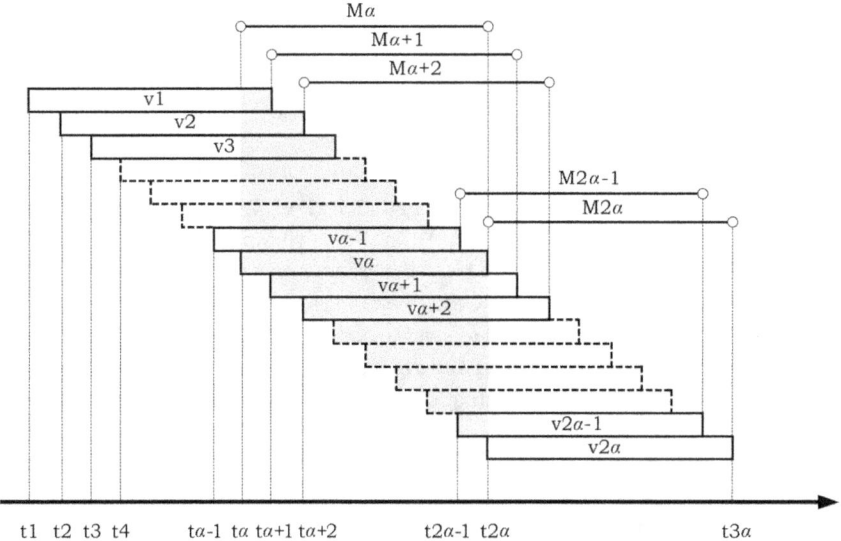

Figure 15.2 The contraction in ECM1. © Ryusuke Okajima.

observing Figure 15.2 vertically, it is clear that, from **tα** to **2tα**, 800 billion vibrations are being experienced simultaneously. The parallelogram-like grey portion we obtain in this way is a single *our moment*, constituted of 800 billion vibrations. If we call this moment **Mα**, the next moment **Mα+1** is the section we can reach by shifting **Mα** slightly to the lower right (although in this figure, vibrations after **v2α** are omitted, the same holds for after **Mα+2**). In what follows, let us designate the model that Figure 15.2 describes the *Extensional Contraction Model 1* (ECM1).

To make our model explicit, let us contrast this with Hirai's contraction model (hereafter, ECM2). Figure 15.3 is the expression by ECM2 of cases where all vibrations in Figure 15.1 are contracted.[7] In Figure 15.3, each moment also contains 800 billion vibrations, but ECM1 and ECM2 represent the inner structure of a single moment and

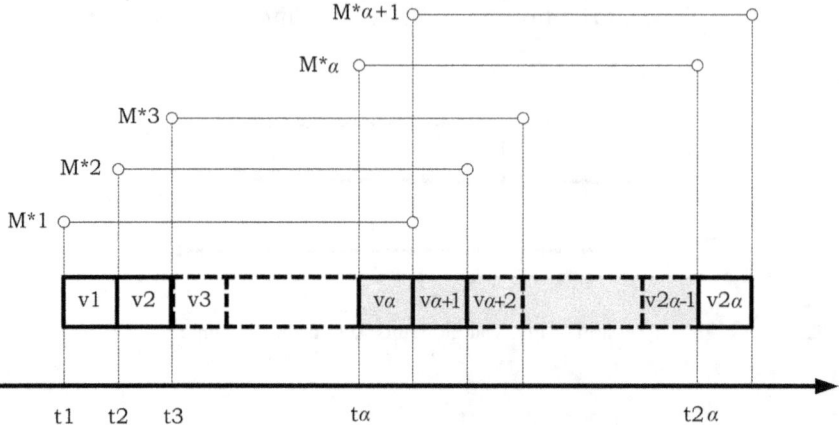

Figure 15.3 The contraction in ECM2. © Ryusuke Okajima.

the connection between moments differently. For ECM2, each of 800 billion vibrations is *wholly* contained in a single moment: in Fig. 15.3, **v1**, **v2**, **v3**, ... and **vα** are all wholly contained in **M*1**. However, for ECM1, only one vibration exists that is wholly contained in each moment: in Figure 15.2, all vibrations other than **vα** are only partially contained in **Mα**. In addition, for ECM2, the *moments overlap each other by sharing their components*, while for ECM1, *components share a single our moment by overlapping each other*: in Figure 15.3, **M*1** and **M*2** overlap each other by sharing all vibrations from **v2** to **vα**, while in Figure 15.2, vibrations from **v1** to **v2α-1** share **Mα** by overlapping each other.

(2) Now, consider the function of condensation. As I already confirmed, for ECM1, 800 billion vibrations are always experienced simultaneously: what is experienced between **tα** and **tα+1** is not the entirety of each vibration but its *parts* that are vertically aligned in Figure 15.2. For accuracy's sake, we denote the parts obtained by dividing nth vibration in the imagination into **α** equal parts, from left to right, **vn(1)**, **vn(2)**, **vn(3)**, ... **vn(α)**: between **tα** and **tα+1**, the vibrations that are experienced simultaneously are **v1(α)**, **v2(α-1)**, **v3(α-2)**, ... **vα(1)**. In that all vibrations from **v1** to **v2α-1** are of the same *type* (red light with a certain frequency), we can see that parts that are equivalent to precisely one vibration are experienced simultaneously in this interval. For this, let **q1**, **q2**, **q3**, ... **qα** be the sensible qualities of each part of the vibration (from top to bottom), and let **Q** be the sensible quality of red light obtained by experiencing all these qualities simultaneously.[8] In short, ECM1 explains the function of condensation with the simultaneous experience of the vertically aligned parts. However, although the terms 'simultaneous' and 'vertically' are used here, this model should not be read in the context of Retentionalism, as no single vibration in Figure 15.2 is found that is reproduced or *retained* in any way. The condensation function is explained here with the simultaneous experience of those parts, but we cannot experience this minimum interval (from **tα** to **tα+1**) in isolation. ECM1 assumes that **Mα** is experienced as an

indivisible thing all at once but also that this structure should be found if we analyse its interior retroactively.

Also, ECM2 is in stark contrast to ECM1 regarding condensation, as the former claims that the sensible quality Q could be constituted by experiencing the parts *horizontally* aligned together, as described in Figure 15.3.

(3) Finally, examining the function of reduction, we note that each moment depicted in Figure 15.1 and each moment in Figure 15.2 is experienced with the same duration: **Mα** exhibits the same duration as that required for the entire experience of **vα**, and this duration is equal to the duration that **ma** has, as it does not change before or after the contraction. So let us denote this duration as **d** for convenience. On the one hand, if no contraction occurs (Figure 15.1), the experience **e** (the material experience of 400 trillion **vs**) naturally exhibits 400 trillion **d** because it consists of 400 trillion **ms**. Conversely, when contraction occurs (Figure 15.2), experience **E** (our experience of 400 trillion **vs**) exhibits only 500 **d** (if we include the time-lag until **M1** is constituted, 501 **d**) because it consists of only 500 **Ms**. In this way, contraction reduces the enormous duration of **e** (400 trillion **d**) to the tiny duration of **E** (500 **d**).

4 Conclusion

Let us summarize our interpretation of the notion of contraction. Bergson's metaphysical premises of direct realism and panpsychism and the results of psychological experiments on perceptual thresholds lead us to understand that the contraction causes individual vibrations to be experienced objectively over a longer time but does not subjectively alter the duration of each vibration.[9] This contraction acts sequentially on an enormous number of successive vibrations, causing them to overlap with each other, resulting in (1) the synthesis of many vibrations into one of our moments, (2) condensation as the vertical effect of the overlap and (3) reduction as its horizontal effect.

This interpretation is inspired by Dainton's overlap model, introducing the idea of overlap into the understanding of the internal structure of our moment and sharing the line of arguments with Hirai's interpretation, which should also be read as an application of Dainton's theory to the notion of contraction. However, each moment in Hirai's model (Figure 15.3) includes the entirety of each vibration, whereas each moment in our model (Figure 15.2), except one, includes each vibration only partially. Also, Hirai describes the formation of mundane sensible quality (like that of red light) through the fusion of horizontally aligned vibrations, while we explain it through the fusion of vertically aligned vibrations. Finally, although this paper places the highest importance on the function of reduction, Hirai does not address this issue.

The purpose of the discussion in this paper is fulfilled above. Let us conclude with a more comprehensive discussion of Bergson's theory of temporal consciousness.[10]

First, it is essential to note that Bergson did not regard the flow of consciousness or, in Dainton's terms, the *succession of experiences*, to take the form of a transition from one content to another. Contemporary theories of temporal consciousness usually adopt this view, taking the example of the succession of do-re-mi tones on the everyday time scale. This way of thinking exactly corresponds to the Jamesian view (James 1950: 243) of the flow (or stream) of consciousness as a 'place of flight' (the hyphens) rather than a 'place of rest' (the do, re or mi). However, as Dainton noted, Bergson criticizes this idea in these terms: 'I see *places of flight* in the *resting-places* themselves, rendered apparently immobile by the fixed gaze of consciousness' (Bergson 2002: 357).[11] In the example adopted in this chapter, Bergson would find a succession of experiences inside a single our moment, such as **M1** or **M2**. As I noted at the beginning of this paper, contemporary theorists tend to deal with the experience of succession only by accepting the *succession of experience* as a given without providing any explanation for this. Even within Bergson's own theoretical framework, contraction alone cannot explain this succession either, as there seems to be no reason in the sensation itself (that the contraction constitutes) to succeed or flow. However, in *Matter and Memory*, we can find some arguments to provide an explanation for the succession of experiences.

> The psychical state, then, that I call 'my present' must be both a perception of the immediate past and a determination of the immediate future. Now, the immediate past, in so far as it is perceived, is, as we shall see, sensation, since every sensation translates a very long succession of elementary vibrations; and the immediate future, in so far as it is being determined, is action or movement. My present, then, is both sensation and movement; and, since my present forms an undivided whole, then, the movement must be linked with the sensation, must prolong it in action. Whence, I conclude that my present consists in a joint system of sensations and movements. My present is, in its essence, sensori-motor.
>
> MM: 177[153]

In addition to sensation, establishing the succession of experiences through the movement of our body is necessary. Here, the movements required are not the actual actions that are already performed but possible actions to be executed, of which we have an awareness at a specific place in the body (to be precise, in the relevant muscles). Although we assume that each of our moments (**M1**, **M2**, etc.) contains only a single element (a sensation derived from external objects), another element of consciousness appears here as well, namely, the movements derived from our body. Therefore, according to Bergson, each moment of our consciousness alone exhibits a definite direction from the immediate past (expressed by the sensation) to the immediate future (expressed by the movement). This means that our moments succeed or flow just as we execute our bodily actions.

Thus, we now possess an explanation of the succession or flow of experiences. However, this discussion does not clarify the *experience of succession* itself (which occurs on a larger time scale). In the do-re-mi sequence, when we take as the moment the last 500th of a second of the mi note, the past that is prior to this is not directly experienced because this past is experienced only when represented in our moment by means of a function of memory called projection (see Section 4 of Hirai's chapter for an overview of this function). In short, Bergson's theory of temporal consciousness is a hybrid theory of Extensionalism and Retentionalism, which allows for a slight temporal extension of the present that is equivalent to the perceptual threshold and provides an explanation for the experience of succession through representation of past events.

This work was supported by JSPS KAKENHI Grant Number JP14J08778 and JP22K12959.

Notes

1. I would like to thank Yasushi Hirai for suggesting improvements to the first version of this paper.
2. Here I refer to E. Pöppel (1988), following Dainton (2000: 170).
3. In *Matter and Memory*, Bergson does not explicitly refer to James on the question of the perception of time. However, given the considerable similarities between *Matter and Memory* and the *Principles of Psychology*, including even a duplication of references to Exner and Boileau-Despréaux's poetry (James 1950: 608; MM: 177[153]) and the distinction between the objective present, as 'strict present'(James 1950: 608) or 'mathematical instant'(MM: 176[152]), and subjective present, as the 'specious present' or 'my present', it is clear that Bergson had James's thinking in mind when discussing the issue of temporal consciousness.
4. Although Bergson himself does not explicitly distinguish between these terms, the following texts allow for such an interpretation (emphasis is mine): '... every concrete perception ... is already a *synthesis*, made by memory, of an infinity of pure perceptions ...' (MM: 238[203]); '... memory *condenses* in each an enormous multiplicity of vibrations which appear to us all at once, although they are successive' (MM: 76-7[73]); '... to perceive consists ... in thus *summing up* [reducing] a very long history'. (MM: 275[233])
5. Note that these figures are taken directly from Bergson's text. In practice, other values may be appropriate.
6. On this point, see Miravète (2012: 290–3) and Miravète (2011: 407).
7. Figure 15.3 presents a reconstruction of depiction A in Hirai's article based on our example. Note that strictly speaking, the figure would not be sufficient for Hirai's purposes because the original depiction does not presuppose the absolute time axis. It is true that in *Time and Free Will*, Bergson does not affirm the reality of an order (usually represented by a timeline) for real-time (duration), as Isashiki points out in his paper (in this collection). However, in my understanding, a change occurs in Bergson's position on this point from *Time and Free Will* to *Matter and Memory* (see, for example, the linear representation of past events in MM: 184[159]). If we adopt the latter perspective, placing the temporal order or linear temporal representation in Bergson's philosophy in a relatively positive way would be possible. I will discuss this issue at another time.
8. Two different possible understandings regarding the sensible qualities of the parts of vibrations are found: (1) they are distinct in quality (q1, for example, is slightly different in quality from q2, q3, etc.); (2) they are exactly same in quality. If we adopt the second interpretation, ECM1 and ECM2 are taking the same view on this point.
9. On this point, I am in complete agreement with Miravète's interpretation of the notion of contraction (Miravète 2011: 290–1) when he says 'moreover, each vibration lasts for a certain period of time, and above all, conserves its duration. These "400 trillion

vibrations" occur in one second in the material world as well as for us. Therefore, the contraction of these vibrations in our state of consciousness in no way reduced the length of time of these vibrations; their duration has not been compressed. The contraction consists in completely different things: it allows to fuse these vibrations in a state of consciousness too small to contain on its surface 400 trillion vibrations or objects distinct from each other' (my translation).

10 For a more detailed discussion, see Okajima (2016, 2018).
11 This is an inadequate understanding of Jamesian theory. According to Girel (2011: 48–56), for James himself, the immobility of the 'place of rest' is only relative, not absolute.

References

Pöppel, E. (1988), *Mindworks: Time and Conscious Experience*, San Diego: Harcourt.
Bergson, H. (1911), *Matter and Memory*, trans. NM. Paul and W. S. Palmer, London: George Allen and Unwin, https://archive.org/details/in.ernet.dli.2015.506437/page/n7
Bergson, H. (2002), *Henri Bergson: Key Writings*, K. Ansell-Pearson and J.Ó. Maoilearca (eds), London: Continuum International Publishing.
Dainton, B. (2000), *Stream of Consciousness: Unity and Continuity in Conscious Experience*, London: Routledge.
Girel, M. (2011), 'Un braconnage impossible: le courant de conscience de William James et la durée réelle de Bergson,' in S. Madelrieux (ed.), *Bergson et James, cent ans après*, 48–56, Paris: Presses universitaires de France.
James, W. (1950), *The Principles of Psychology*, vol. 1, New York: Dover Publications.
Miravète, S. (2011), *La durée bergsonienne comme nombre et comme morale*, doctoral thesis, Toulouse: University of Toulouse II.
Miravète, S. (2012), 'La durée bergsonienne comme nombre special', *Annales bergsoniennes* 5: 401–18.
Okajima, R. (2016), 'Theory of Temporal Consciousness of Bergson: From *Time and Free Will* to *Matter and Memory*', *Journal for the Philosophical Movements in Tsukuba*, 24: 25–38.
Okajima, R. (2018), 'On the Notion of "My Present" in *Matter and Memory* of Bergson', *Annual Review of the Phenomenological Association of Japan*. 34: 93–100.

16

We Bergsonians: The Kyoto Manifesto

Elie During and Paul-Antoine Miguel

'We Bergsonians...'

Of the philosophers working today, who deserves to be regarded as a true Bergsonian? Conferences and seminars devoted to 'Bergson studies' may not be the best testing ground. When Aristotle paid tribute to his master in the famous section of the *Metaphysics* on the doctrine of forms ('We, Platonists...'), it was to develop his ideas in new directions. The result, as we all know, was a different philosophical system entirely. It would be astonishing if re-thinking the doctrines of *durée*, becoming and genuine novelty were not similarly capable of yielding new insights and pointing to new directions beyond Bergson's work: not merely a *return* to Bergson, a *neo*-Bergsonism, but a philosophy for our own times.

A philosophy, however, cannot be instituted by decree; it needs to be actually created. Now it may well be possible to be a Bergsonian despite Bergson, and – at least up to a point – against Bergson, but if the distinguishing trait of Bergsonism is its methodology, we first need to ensure that this methodology continues to be of genuine use. We need to explore what this methodology can deliver today, with regard to the problems that *we* are currently confronted with. In this regard it is obvious that Bergson scholars aren't themselves always very Bergsonian when it comes to the concrete forms of philosophical inquiry. As Gilson aptly put it: 'The true Bergsonians are not those who merely repeat Bergson's conclusions. Rather, they are those who – following his example – make these conclusions their own, and in different areas succeed in doing something analogous to what Bergson did.'[1]

We, Bergsonians, have read and re-read Bergson; we have studied the complex ways in which his philosophy has been received. We have defended him against his detractors; we have corrected misunderstandings, provided the overlooked context of his *oeuvre*, and felt the singularity and force of his theses, the subtlety of his way of thinking: demanding and 'difficult', as Bergson himself acknowledged, misleading in its apparent informality ('How on Earth did anyone miss that?' he wonders) All this was necessary. But now is not the time to give a second youth to Bergson's *'philosophie nouvelle'*. Bergson is already amongst us, and he is not lacking in friends. The question we find ourselves confronted by is how best to harness the impetus of his philosophy, even if this involves directing it along new lines.

Bergsonism has been interpreted in various ways: the point is to change it and put it to work in the context which is manifestly very different from Bergson's own.

We are setting out the case for *an expanded Bergsonism*.

Expanded Bergsonism

'Expanded' in what sense? First and foremost, our Bergsonism must engage fully with the sciences of today. It is important to remember that Bergson himself was continuously stimulated and spurred on by the sciences of his time. His method risks amounting to no more than an idle play of concepts if it is confined to purely intra-philosophical issues. What is the point of reformulating the great metaphysical questions – determinism and contingency, matter and mind, monism etc. – in terms of *durée*? Metaphysics is never conducted in a vacuum; it is never purely *a priori*. Bergson insists that intuition is a form of reflection, but reflection on what? Invoking the distinctive character of lived experience is of little help if it is not in constant touch with the rich experiential content elaborated in the worlds of science and technology. Bergson wanted us to engage with the full-blown spectrum of experience – 'integral experience', as he calls it. Admittedly, scientific experience is essentially committed to a symbolic framing of the given, but it would be strange indeed to exclude from our inquiries such an essential part of our cognitive activity, or to confine it to the margins of the 'ordinary'.

The necessity of once again putting Bergsonism into direct confrontation with the living sciences – science in the making – illustrates a more general requirement. What we need is not a new commentary, it is a new research programme: one adapted both to the problems of today and the constellation of questions that Bergsonians have traditionally focused on, but which permits deviation from methodological orthodoxy if and when the need arises.

To successfully carry out a project of this sort, a project which takes Bergson as its point of departure, it isn't necessary to speak his language – on occasion, it may even be preferable not to. To re-iterate: the task at hand is not to read or re-read Bergson, it's a matter of *translating* him. Translating Bergson requires us to escape the stranglehold of his own words. We need to assess how well his claims fare in new and different contexts and frameworks. This stress-testing will bring Bergson into contact with different philosophical traditions, with their own terminologies. Yes, we must continue to confront Bergson with phenomenology and analytic philosophy, method against method. We need to imagine – why not? – an analytic Bergson, a *philosophically* clean-shaven Bergson, as Deleuze said about Marx. Such experiments are not intended for the perverse pleasure of watching Bergson turn in his grave. The aim is to bring the distinctive features of a philosophical position into clear view, and so bring the relevant domains of research into proper perspective – while in each case isolating the critical connection between philosophy and the empirical sciences.[2]

There is an immediate advantage to adopting a *broader, expanded Bergsonism*: those who find much to admire in Bergson, but who are unable to count themselves as his disciples, need no longer suffer from a guilty conscience. It is possible to be a Bergsonian without giving any thought to safeguarding an orthodoxy from its detractors. To extend one of his intuitions in a different direction, or find new avenues of attack, nothing forbids us from reconstructing his doctrines, or finding novel ways of recombining his concepts. Similarly, there is nothing to prevent our transforming or deepening his analyses by bringing them into contact with other philosophical programmes, provided they intersect Bergsonian thought by virtue of having certain problems in common. The expanded Bergsonism advocated here will thus be a *refracted* Bergsonism. This comes at a price. The risk is to run on occasions into dead-ends or blind alleys. It may turn out that we have been in pursuit of a mirage, a merely virtual bergsonism.[3] However, this, perhaps, is the price of all real experimentation. As Jacques Rancière said with regard to Deleuze: 'The strength of every strong thought is also its ability to arrange its *aporia* itself, the point where it can no longer pass' (Rancière 2004: 164).

In the paragraphs which follow we outline the general shape of such an expanded Bergsonism, while reviewing some of the issues it must confront, as well as the specific areas where we believe new research should be concentrated.

A non-speculative empiricism: following 'lines of facts'

Let's start by marking one vitally important point. Bergson subscribed, as is familiar, to a 'true empiricism'. He spoke of a 'positive metaphysics' in order to suggest that the philosophical elaboration of experience involves more than a merely verbal reference to the whole of experience. By contrast with speculative metaphysics, positive metaphysics develops in close – and critical – connection with the symbolic frameworks devised by those whose primary aim is precisely to come to terms with the facts.

The 'immediate given' has sometimes been overlooked or misunderstood: there is no phenomenologically pure experience that can be taken as philosophically foundational. Even the most primitive forms of experience are compromised with forms of objectivity originating in our cognitive activity: the primary manifestations of 'intelligence' are as much a given as the revelations expected from intuition. If it were not so, it would be impossible to understand why Bergson attempted to forge the critical tools he did *right from the start*. It is critical that we distinguish and disentangle space from *durée*, but we have no choice but to start with this state of entanglement. Bergson's pure *durée* is not simply given; it requires effort, it must be conquered. We cannot do otherwise than starting from where we stand, in the midst of confusion. The 'turn of experience' is a vanishing point, it will not yield 'first principles' for our philosophical guidance. The philosophical method of intuition cannot be *a priori*, nor does it take the form of a transcendental structure in the classical sense.

This isn't to say that that there is no philosophical grammar, and hence that Bergson's spiritualist empiricism reduces to purely critical manoeuvres performed in the vicinity of the natural sciences. The grammar is there, but it is sometimes concealed, or only implicit, in the texts. It's a *constructive* grammar, rather than one that is given, or deduced. It must be understood in relation to a metaphysical imagination which, retrospectively, provides us with something akin to a sketch: a nascent vision elaborated in the guise of a schema for reality. It is only by resolving what is blurry in the picture or indeterminate in the schema that metaphysical precision will be attained. The only way metaphysics can escape metaphors and be taken literally is by assuming this figurative condition.[4] This is the source of the difficulty of Bergson's language, and all the ambiguity of his allegedly 'imagistic' prose. An image 'isn't a mere ornament, it's a suggestion of representations which, in order to be fruitful, should suggest lines of research',[5] he writes. This is why several images are usually needed: if possible very different from one another, where these differences reveal the inadequacies any single image might possess. An image offers a line of sight, and in that respect it fulfils exactly the same role as the lines of facts. The latter too must work together – they only reveal their full significance when they end up intersecting with one another.[6]

Thus, phenomena which are at first sight as disparate as paramnesia (the sense of déjà vu), the 'panoramic vision' of the dying, or the stubborn persistence of a 'word at the tip of one's tongue', can all point to the existence of a single distinctive mode of being: an immemorial past that produces and conserves itself as reality is itself being made. The facts are not only 'stubborn', as Lenin put it: they are above all disjointed and scattered. That is why we need to re-arrange them in accord with the 'lines of facts'. The difficulty is to demonstrate – via tentative exercises of trial and error, while always taking care not to lose sight of the heterogeneity of the given – to make these diverse lines of facts converge on the same intended target. If it were not supported by this patient work of focusing, the integral conservation of the past would remain nothing more than an obscure metaphor. In all cases it is necessary to view the question under consideration from multiple strategic vantage points: 'Intuition isn't an inspiration that has dropped down from the heavens; it's an approach which involves leaping to the centre, after having taken sightings from all sides' (Adolphe 1951: 4).

Ontological commitment and critical distance: the connection with the sciences

If Bergson's thought has never transformed itself into a neo-Bergsonism, a *spirit of Bergsonism* has nonetheless had an impact on the entirety of contemporary French philosophy, from Canguilhem to Deleuze, passing through Simondon, Ruyer, Merleau-Ponty and Foucault, to mention only a few. Its distinctive features include recognizing the importance that should be given to the lines of facts, and to the empirical sciences which provide us with a detailed understanding of them.

In this regard we must note two points that continue to be overlooked by the mainstream reception of Bergson's ideas.

1. Contrary to what has often been assumed, for Bergson scientific knowledge is not necessarily artificial, a mere play with abstract symbols. Science would not be as effective as it is if it did not carry ontological significance: in its own way, 'science reaches the absolute'. Even mathematics, in certain respects, 'is by no means just a game, but a real point of contact with the absolute'.[7] 'I would like to know if there exists, among the contemporary conceptions of science, a theory which gives to the empirical sciences a higher status.'[8]

2. As conceived by Bergson, intuition is not the perfectly precise instrument that it has sometimes been taken to be, a kind of laser beam by which we are able to penetrate through to the ground level of reality, where we directly apprehend becoming. Intuition provides the driving force, but left to itself its deliverances remain essentially vague. Intuition only becomes precise when its scope is focussed and tightened.[9] It needs, as we have just seen, to follow the lines of facts, and be transposed into symbolic-conceptual forms.[10]

There is also a practical necessity. In order to be communicated at all, intuition requires intelligence, it has to 'ride on the back of ideas'.[11] But Bergson takes a further step:

> As soon as we have intuitively apprehended the truth, our intelligence corrects itself, and gives intellectual form to its error. It has received a suggestion/hint, it provides in its turn control. As the divers will touch in the depths of the ocean the shipwreck that the aviator has spotted from high in the air , so our intelligence – immersed as it is in a conceptual environment – will analytically verify, point by point, what has been apprehended in a synthetic, supra-intellectual manner.
> Bergson 2007a: 67

And better:

> The *intuition* that I have spoken of can only enter into play after one has studied, deepened, and often even expanded upon the empirical evidence that positive science has gathered on a certain issue.[12]

Science can reach the absolute, and intuition has need of science, not only to be communicated and conveyed, but simply in order to function. For these two reasons, science is a good deal more than a domain in which the Bergsonian methodology can be applied. Above all, the role of science is not confined to that of the bogeyman that possesses only functional, 'tool-making' intelligence, incapable of understanding life, that 'spatializes' and distorts everything it touches. This picture is all too familiar in the textbook version of Bergsonism, but it is utterly misguided. Science is not there simply as a foil that one would set up to promote a general critical methodology aimed at recovering the vital, pre-reflective aspects of reality. Rather science is the necessary precondition for the effectiveness of the method in question. Bergson's engagement with science goes beyond both the hermeneutic mission of interpreting scientific works and elucidating the ontologies implicit in scientific theorizing: philosophy, Bergson maintains, must 'mould itself upon science'. However, this intimacy indicates a true complementarity, rather than any subordination of one to the other, and this in turn presupposes that intuition introduces a *critical distance* – rather than a rift – with regard to the sciences, albeit in a very different manner than phenomenology or analytic philosophy.

This is arguably the most important feature of 'intuition as method', to follow in Deleuze's steps. Intuition manifests itself initially by its capacity to refuse certain readymade intellectual schemas. Accordingly, the Bergsonian methodology includes a crucial analytical aspect, which is manifested in the dialectical manoeuvres aimed at identifying and eliminating inadequate concepts and erroneous syntheses. Science is concerned to the extent that it sometimes yields to misleading analogies and over-hasty generalizations. These can be prevented and corrected, with the aim of bringing into the open the inconsistencies of the (often unconscious) metaphysical assumptions underlying scientific discourse. The proper use of intuition, Bergson believes, will open new lines of questioning, and sometimes lead to factual discoveries which may in turn lead to the reform of certain habits of thought shared by working scientists. To sum up, philosophical interventionism is the price one must pay for acknowledging the ontological weight of scientific discourse. Our contention is that a Bergsonism which had nothing to contribute to the improvement of science would not be worth an hour's effort.[13]

That said, the proper role of intuition does not extend to supplying science with metaphysical 'foundations': science manages to function perfectly well without these. Nor does intuition aim at panoramic 'conceptions of the world', as if the task of philosophy was to supply science with broader and bolder syntheses, taking its theories to ever higher levels of generality along the same directions. This would condemn philosophy, which can only ever engage with the sciences as they are at a particular moment in time, to always lag behind on the path of scientific progress.

Bergson has thus tried to connect metaphysics with the sciences in an entirely new way. He started from the assumption that philosophical questions enjoy an autonomous existence – something that is always difficult to convey to scientists who are as likely to forge their own 'philosophy' as they are to repudiate all philosophy as entirely useless. By demanding further refinements, distinctions and differentiations, however, science invariably renders these same questions *more precise* than they would otherwise be.

> I see in the metaphysics that is to come an empirical science of sorts, one that is cumulative, but also constrained – in the manner of other sciences – to treat the results to which its attentive study of the real has led it, as being only ever provisional.[14]

Given this, in which directions should our efforts be directed? What are the questions which should be occupying us today? Which lines of facts are ripe for philosophical scrutiny?

> Where will we be led? No one knows. No one can even tell us which science will uncover the new problems. It could even be a science with which we are today wholly unfamiliar.
>
> Bergson 2007a: 72

Neither reductionism nor emergentism

Let's try a few guesses anyhow. By way of a concrete example, consider living organisms, or organizations. The real character of Bergsonian vitalism – often obscured by an overly strong fixation on the doctrine of the *élan*, or by an overly simplistic opposing of mechanism with finalism – has been rediscovered. From a critical perspective, the point to keep in view is that the Bergsonian approach provides an alternative to conceptions of nature that are either *reductively* physicalist and *emergentist* and *hierarchical* (with regard to both nature itself, and the relationships between nature and culture). Emergentism can be developed in terms of levels of reality, or levels of explanation (with higher-level properties 'supervening' on more fundamental properties). From the Bergsonian perspective there is a plurality – no doubt irreducible – of modes of existence, but this plurality is not distributed in layers corresponding to increasing levels of complexity, crowned by the domain of consciousness; reality simply isn't stratified in such a straightforward manner, because genuine complexity cannot be the product of a mechanistic combination of elemental bricks of reality. As Ruyer puts it, 'the "new" is being formed at each instant everywhere, but there are no superimposed layers ... each bringing its characteristic "novum"' (Ruyer 2012: 272–3). (It goes without saying that Bergsonism also rescues us from more sophisticated forms of reductionism or regionalism, such as Husserl's philosophy of consciousness, or the ontological phenomenology of *Dasein*.)

The Bergsonian approach to the nature of life starts from the idea that we are always already immersed in entangled levels of *durée*, not (as Deleuze would have it) in a 'plane of immanence', nor even in a plurality of such planes, but rather in a *redoubling* of immanence. First of all because, as Frédéric Worms has noted, there really are *two* meanings of life.[15] Underlying the distinctive doctrines of *Creative Evolution* is a *cosmological turn* in Bergson's thought, which accounts for the fact that to experience the immanence of durée it is necessary to be able – simultaneously – to experience the 'durée that is immanent in the whole universe', that is the cosmic consciousness to which our own consciousness is connected by innumerable links of sympathy. The paradigmatic example is the well-known 'lump of sugar': its melting participates in the absolute in *two* respects, because in addition to bringing lived duration in the focus of attentive awareness, it also bears the weight of the entire cosmic process. On the one hand, the melting is the living symbol of my own impatient waiting; on the other, my impatience is pitched against the whole spectrum of intertwined durations that make up my environment: I have no choice but to wait, and the lump of sugar is the living symbol of the unfolding universe.[16]

These redoublings make it possible for particular episodes of *durée* to constitute themselves, not only by enfolding or contracting more or less intense instances of duration (this is how Bergson accounts for our phenomenological *situation* within the world), but also, at the same time – and to different degrees – by extrapolation or projection into other episodes of *durée* which in turn encompass them (this is the theme, less obvious but clearly present, of *perspective*, to which we will be returning). Bergson introduces fundamental distinctions which help us orient ourselves within this entangled domain governed by reciprocal causality: if we are considering things statically, the relevant distinction is between *the closed* and *the open*; from a dynamic perspective the relevant distinction is between *iterative repetition* and *creative repetition*.

We believe the entangled levels of *durée* are a fundamental characteristic of the living, one of the basic parameters determining biological organization. The concept of redoubled immanence takes a more explicit form in Simondon's work, with his claim that biological individuation should be construed as taking place in a 'theatre' in which it plays the role of an 'actor'. The same intuition is at work in Bailly, Longo and Montévil, when they urge that we conceive of the transition from the physical to the biological in terms of an organism which lives and develops, repairs and reproduces itself in a *space of extended criticality* which exists in addition to the standard range of physical properties (Longo and Montévil 2014).

The virtual and the possible

From a Bergsonian perspective, the virtual is real, but the virtual is not simply the possible. Life and memory, to the extent that they are virtual and never cease virtualizing themselves, mean we must re-evaluate the character of the modal relationship between the possible and the actual, between *what is* and what *might be*. In other words, to take proper account of the nature of the virtuality which underpins real-world processes, we cannot consider what might be as simply a form of what is – and hence as something amenable to direct quantification and calculation.

Suppose, for example, that we take the states of a biological system to exist in space of possibilities which includes both the random and the probable. In so doing we would not be in a position to explain the radical novelty and unpredictability which, in reality, are characteristic of such a system. The evolution of life – and the same goes for human memory – creates at each instant *new possibilities*. Evolution acts on and alters, in real time, the space of possibilities which (supposedly) explains evolution itself. We are thus led to distinguish the concepts of what 'is different' and what 'might be', between what is simply *possible*, and what is *truly contingent*, and by nature unpredictable. These conceptual clarifications will no doubt have an impact in domains such as decision theory and the philosophy of action.

A broader logic

Bergsonian ontology requires – as with Simondon, albeit in a somewhat different way – a broadening of logic. This broadening does not involve our putting into question the principle of non-contradiction, but it does require us to reject the *principle of the excluded middle*: Not not-A and A are no longer equivalent.

This brings us back to Bergson's views on the 'two orders'. Matter is not merely negative, it is not the reverse image of life, it does possess a form of *durée*, different in nature from that of living things. On the one hand, the *durée* of matter is characterized simply and solely by repetition, whereas the *durée* of living things is characterized by invention, creation and self-overcoming (at least with respect to their purely material nature). On the other hand, it remains the case that matter belongs to life, and is somehow folded within it. We here run into the issue of emergence once again. Is organization something which emerges from a primitive physical substrate, or is it the other way around? Isn't the physical best understood as the neutralization or slowing down of the interactions that are characteristic of the biological realm? This provides us with a fine example of a non-conservative extension: to comprehend how the physical can go beyond its own order it is necessary also to understand that the alleged physical basis was itself the product of an idealization.

Similarly, for Canguilhem an illness isn't nothingness, but rather another mode of life – or (if one prefers) an illness is a deviant physiology, which also means it can't be reduced to the merely physiological. Inversely, life is a 'dynamic polarity', which means that life isn't simply the opposite of death, nor health the opposite of illness: rather there is a pathological dimension inherent in life itself – death is to be found within life and nowhere else.

Again, the issues are clearer and more explicit in Simondon. 'To understand individuation,' he writes, 'one needs to conceive of being not as substance or matter or form, but as a system that is under tension, oversaturated, that exists over and above the level of unity, which doesn't consist solely in and of itself, and which is therefore not constrained by the law of the excluded middle' (Simondon 2005: 25) Which amounts to saying that becoming itself is not nothing, or equivalently, that being isn't whole. Just as becoming has a mode of being, so too does being undergo becoming. Being, as Simondon points out, can be out of phase with itself.[17]

It is only by rejecting the excluded middle that we can hope to be in a position to understand how it can be that in the order of things, it is the virtual rather than the actual which comes first. If the virtual is *otherness*, the actual is that which is *other* than otherness.

Presentism and eternalism

It is from this basis that one should consider the critical relevance of Bergson to contemporary debates concerning the metaphysics of time. In the analytical world we find 'presentists' and 'eternalists' doing battle over the domain that should be given to the verb 'to exist'. (Should it be confined to only the present? Or to the past as well, or perhaps even the future?) It is clear, however, that their disagreement rests on a presupposition that is common to both camps, but which passes largely unnoticed. It is being assumed, in effect, that the past and the present are fundamentally homogeneous in nature: that the past differs from the present only by virtue of *no longer being* present, and hence that the past and present have exactly the same *kind* of contents, which is apparent in the fact that, from the point of view of ontological calibration, they can only be differentiated from one another modally. Once this assumption is made, we still need to explain how the concepts of past and present nevertheless apply to successive states of affairs that do not differ in kind. Some appeal to a mysterious 'passage' of time (which one can try to elucidate via the analysis of various 'models' of the flux of consciousness.[18]) Others appeal instead to perspectival notions compatible with the view that reality in itself is perspective-free (hence the frequent appeal to the indexical character of 'now': what counts now as 'now' is dependent upon our contingent location along the timeline). If, however, one views the present and the past as fundamentally different *in kind* – as different, according to Bergson, as the actual and the virtual – then the metaphysical landscape is entirely transformed and the philosophy of time must be grounded on new ontological foundations. To achieve this, it is interesting to filter Bergsonian themes through the analytic instruments provided by current debates in the philosophy of time. The quarrels between 'tensers' and 'detensers', proponents of A-time and B-time, should not be dismissed as moot. Even pseudo-problems can be of use when it comes to making Bergonians ideas more precise. For it is not enough to gesture towards the virtual, or to claim that space-like models of time give up genuine becoming from the start. Gnomic formulations will remain opaque no matter how frequently they are repeated – such is the formidable claim that the past conserves itself integrally 'in itself' ... Here again, philosophical debates out new lines of facts, following, if need be, the latest developments in neuroscience.

Re-examining the relation between duration and space

We said above that we shouldn't hesitate to loosen the stranglehold of Bergson's language and terminology. Let's start with *durée*, Bergsonism's central concept. In the literature it has generally been reduced to nothing more than a qualitative or heterogeneous multiplicity, entirely lacking any numerical dimension. As established by several recent studies, this idea has had its day.[19] In fact, *durée* and number are not entirely foreign to one another: *durée* can be viewed as 'an obscure number', 'a special number' – i.e. non-spatial in nature. This idea, if properly developed, would lead us to a qualitative arithmetic, one that is capable of describing physical phenomena without obliterating the temporal aspect inherent in their becoming. For after all, what is measurement if not a way of coming to terms with the internal rhythms of change?

Among the other taboo words for Bergsonians, 'spatialization' is not the least problematic. What is space? The critical role the category plays in Bergsonian thought is clear enough, and there's no need to return to the relevant points here. However, how do things stand with regard to the diverse plurality of spaces, both concrete and abstract, that are associated with the lives of humans and animals, as well as with scientific and artistic activities? Are there lines of facts here which might be of interest to philosophers? Bergson famously criticized the metaphysical assumptions behind the adoption of one particular kind of abstract space in connection with relativity theory: the four-dimensional Minkowski spacetime, with its characteristic pseudo-Euclidean signature. He took this construction to be simply an extension of the kind of geometrical representations of movement that are frequently to be found elsewhere, e.g. in the ordinary use of curves and graphs depicting change-over-time. Would he have had the same to say about configuration or phase spaces in other areas of contemporary physics? If the spatial mode of existence is indeed the ideal limit to which matter (*qua* extension) tends by its own nature, how should this metaphysical thesis be situated with respect to the diversity of geometrical constructions to be found in contemporary mathematics: not only n-dimensional non-Euclidean spaces, but complex spaces, discrete spaces, and the like? To turn to a related topic, what status should we give to set theory? Is it merely a natural extension of the Bergsonian notion

– quite new at the time—that mathematical objects should be construed as *systems of relations* between elementary objects (points, instants)? Lastly, what should we say about the purely topological properties of mathematical spaces? Bergson's critique of measurement was originally developed in connection with very basic examples of the kind one may find in the Transcendental Aesthetics (tracing a line, building a triangle in Euclidean space, counting successive moments . . .). Does the critique apply equally well to the more elaborate, but nonetheless effective and useful modes of spatiality developed in the wake of modern geometry and physics?

There are a thousand reasons for remaining less than satisfied by the Bergsonian treatment of the relationship between *durée* and space. Suppose we take seriously the idea that our intelligence is not confined simply to the pragmatic task of manipulating the physical things by hand or manufacturing tools for doing so, but is ontologically committed with the genesis of materiality, as Bergson claims. Suppose, in addition, that the universe as a whole is continuously undergoing becoming, as Bergson often insists, and that space is nothing more than matter (and hence, duration) in its most relaxed state, i.e. an ideal limit for which geometrical concepts provide partial yet effective templates. Then it is clear that *real space* – let's introduce the term by way of symmetry with the 'real time' discussed earlier – can have a content that is far richer than that allowed in *Données immédiates*, which is in fact mainly concerned with the intellectual schemes underlying the representation of space as numeric multiplicity. If there is such a thing as real spatiality embedded in the unfolding universe, surely intelligence must have a grip on it. It is only the habits reinforced by the endless reliance on 'useful conventions', to use Poincaré's expression, that condemn us to eternally orbit the circles of Euclidean geometry. It remains to be seen whether these richer and more diverse modes of spatiality can be accommodated in the intensive schema elaborated from *Matière et Mémoire* onwards, with its degrees of duration and variably contracted rhythms.

The problem of co-existence: the local and the global

Contrary to what Bergson himself sometimes suggested, the question of the nature of space lies at the heart of his debate with Einstein. One reason for this is the simple fact that the concept of simultaneity is occupying centre stage – though it should be noted that in his discussion of Einstein, Bergson was operating with a conception of simultaneity that is altogether different from the conception that can be found in his earliest works, where it is defined as the mere intersection of time with space. There is a second reason: in developing his concept of 'real time' in *Durée et Simultanéité* Bergson was intending to project durée onto a particular aspect of *measured* time. While this may come as a surprise to some, there can be no doubt at all that real time – the figure of *durée* which comes to the fore in *Durée et Simultanéité* – is a kind of time that is inherently measurable and quantifiable. The distinction between real and fictitious time (e.g. symbolic, or purely mathematical representations of the temporal) runs within measured time itself. If it were otherwise Bergson would never have dreamed of resisting the *metrical* consequences of the famous 'twins paradox' of relativity theory attributed to Langevin.[20] What Bergson contests, clearly wrongly – as confirmed by all the relevant lines of facts – is the thesis that the twin who is accelerated in his spaceship can give a different *measure* of the *durée* that has elapsed during his trip, a *durée* that is thus (allegedly) substantially identical, but also metrically equal, to the *durée* experienced by his brother who remains on Earth. In more precise terms, Bergson holds that there is only a single interval of 'real time' between the moment when the twins separate, and the moment when they are reunited. It is clear, however, that what is at stake here is not *durée* in general, but only the particular times measured by conscious human observers, who are able to plot the flux of their lived experience against the reference frame defined by their measuring equipment.

What is of interest to us in Bergson's entangled discussion of this typically relativistic setting is the more general problem regarding the framing of time. The point is that real time isn't necessarily lived time: Bergson only requires that time *can* be lived through. In contrast, it's essential to real time that it should be actually measured.[21] Consequently, it is not enough to remind us that *durée* in its pure state is by its nature unmeasurable or unquantifiable. We must accept that lived *durée* can, in a manner of speaking, take leave of itself and adopt an observation point in the domain of the measurable, so as to give a *perspective* on other *durées*, and so recognize that their flows unfold simultaneously. For there is no room for doubt: the twins are, in some real sense, contemporaries. To accommodate this fact, we have to accept that the 'proper times' that can be found in distinct spatial locations are in some way commensurable. This is the theoretical setting that Bergson wishes to consider. Are relative measures of elapsed time enough to capture the metaphysical import of the thick notion of simultaneity that Bergson has in mind (i.e. a simultaneity not confined to zero-duration instants)? The difficulty is that the 'proper time' of the equations of physics – the quantity that is guaranteed to be invariant through changes in frame of reference – is a time which is completely lacking in global significance: it is a strictly *local* determination of time, bearing no perspective on other times. Mathematically speaking, it primarily functions as a parameter. It is in some respects less artificial than the coordinate times linked to particular inertial reference frames, but there is nothing intrinsically temporal about it. As Piaget would put it, proper time is nothing more than 'spatial succession'. Unless that is, a sort of 'reality transfusion' occurs whenever we consciously think about it: then, the worldlines traced out by the occupants of four-dimensional Minkowski spacetime become infused with *durée* deriving from our own consciousness, with the resulting illusion that the entire scene can be re-animated and set into motion despite the lack of an overall time frame. Truly speaking, however, what we are dealing with here are purely local and *a priori* disjointed measures of material *durée*. These measures exist alongside the *basso continuo* of the lived durée of conscious observers – a *durée* characterized by the degree of tension typical of humankind – without any straightforward notion of real simultaneity between flows. Any attempt to uncover from these ingredients a genuinely temporal unity of the material universe is bound to fail. Anyone attempting cosmological unification on such a basis would run the risk of indulging in psychological metaphors (a super-human consciousness sweeping across the universe ...), at the risk of being swiftly undermined by scientific objections – not least those of Einstein, who, it will be recalled, refused to admit the existence of a philosophical time, distinct from the time of the physicist and the time of the psychologist.[22] Above all, what is still lacking in the way the problem is framed by physicist and most philosophers of physics is a clear conception of temporal perspective, one which acknowledges the possibility of effectively extending time through space, via the operation of measurement. What we still need, in other words, is a conception of real space – or if one prefers, real simultaneity – that is fitted to address the cosmological challenge.

At the very least, Bergson managed to point his finger in the direction of the principal problem: finding a mode of cosmological connection which accommodates the legitimate demands of *situation* (*context*) and *perspective*, with determinations of time that are *global* as much as *local*. His confrontation with Einstein's special theory of relativity led him to formulate an issue that is crucial to our understanding of the nature of physical time: the simultaneity or coexistence of spatially separated flows (or *fluxes*) of duration, a coexistence that is irreducibly temporal and spatial, and which in no way undermines the evidence suggesting that instantaneous simultaneity (the punctual 'now') is by necessity relative to the choice of a particular reference frame (i.e. a system of coordinates extending through the totality of space).

Admittedly, the familiar Bergsonian distinction between *durée* and space, qualitative multiplicity and quantitative multiplicity, isn't a great help here: the best it can do is providing us with a name for our problem. The problem of simultaneity – of real simultaneity, envisaged *sub specie durationis* – should be formulable in terms of space-time. It is unfortunate that physicists continue to think they are doing philosophers a favour in conceding to them that the distinction between lived time ('psychological time') and objective time ('clock time') still has relevance and validity. For this reason, and not withstanding their prominent role in mainstream Bergsonian scholarship, we propose to banish all use of expressions such as these, once and for all. It is now apparent that everything hinges – as in reality it has done since the start of the debate – on measured time. Giving a privileged status to psychological time – as if it were a valuable treasure which must be preserved from the threat of our inherent spatializing tendencies – assists not at all in solving the problem. Langevin once proposed translating 'lived time' by 'proper time', as if it were a matter of finding a common lexicon. This concession to the philosophers was meant as a conciliatory measure, but in the end it did more harm than good. In the absence of a single temporal metric extending over the whole of spacetime, any common measure of time at a distance is arbitrarily bound to a particular reference frame. The fact that similar temporal units are used across space produces an illusion of temporal homogeneity. The truth is that in space-time one only finds *spatio*-temporal homogeneity: the intervals of proper time corresponding to different spatio-temporal trajectories remain in a sense incommensurable, which lends support to the notion that time is fundamentally and irreparably disunified. The paradox of the twins, who are somehow contemporary despite their diverging temporal experiences, embodies this general issue. Einstein was candid enough to admit that he did not see any genuine problem there, but the issue has recently returned in the guise of a proliferation of *non-standard definitions of simultaneity*. If much of the philosophical debates largely ended up focusing on the issue of whether such definitions are conventional or non-conventional, it's interesting to note that metaphysicians in the analytic tradition have recently been exploring the question while not confining themselves to strictly epistemological considerations. These philosophers have started developing *regional* models of coexistence (featuring an extended, non-pointlike conception of location) – these have the advantage of being directly derivable from the *topological* structure of Minkowski space-time, thereby giving access to the deeper features of real space.

Between confabulation and creation: living and thinking with science

By way of conclusion, let's return to the relationship between philosophy and the sciences. Bergson laid the foundation stones of his methodology, and undertook his early research in a climate that was still largely dominated by a positivistic faith in the elucidatory powers of mechanistic modes of thought. With the advent of the twentieth century, he was alert – as were others – to what would soon become known as the 'crisis of the sciences', a crisis to which he would himself contribute: for example, by popularizing a pragmatist conception of scientific theories, which were to be viewed as intellectual tools, designed to capture only the measurable aspects of physical reality. However, there is no getting away from this fact: much of what Bergson had to say on these matters needs revisiting in the light of the revolutions which occurred in the first third of the twentieth century, revolutions that so profoundly disrupted the scientific image of the world. Needless to say, we are thinking here of general relativity, about which Bergson said practically nothing, but also of quantum mechanics, molecular biology and information theory – without even mentioning the foundational advances made in logico-mathematical fields. These revolutions have led to a revision of the categorical frameworks on which the philosophy of the natural sciences. They have also impacted our understanding of the significance and scope of scientific theories in general. Their sheer diversity greatly complicates the 'grand unification' project, which is based on the idea that nature is a totality governed by universal laws. If our expanded Bergsonism is to avoid exhausting itself fighting straw men – in the form of a 'mechanistic' science which had already been superseded in our grandparents' day – it is crucial that we take proper account of these recent developments.[23]

We noted earlier that a positive metaphysics must establish a 'critical distance' with respect to scientific theories, where this goes beyond elucidating their epistemological or metaphysical foundations. The need for such a distance derives, in part, from the character of the sciences themselves, of which we now have a better understanding. To put it succinctly: the ontological claims made by a scientific theory are genuine, but invariably *limited*. Although a theory will provide us with a description of reality, it does so only in a certain manner, under a particular perspective, and it will not be able to supply a full justification for so doing. Hume was the first to make this point: a scientific theory will always be incapable of supplying its own foundation or justification. Thanks to the work of Gödel on the incompleteness theorems, we can say the same today about the formal sciences.

From a Bergsonian perspective, there is nothing surprising here: *even mathematical proofs are, ultimately, nothing more than formalized gestures*. As Guiseppe Longo has established, as well as rules of inference, they invariably employ constructive principles (Bailly and Long 2006). This 'deterritorialization' of science (as we might put it) is not a product of the content of its theories, but is due instead to their essentially pragmatic or practical nature. We find here, illuminated from yet another angle, the redoubling of immanence. A scientific theory tells us, at the level of thought, of a world other than thought; it concerns the very same physical and biological world to which we are ourselves anchored by the demands of action. As a consequence, science brings into play what we can call an *enlarged experience*, or alternatively, a *decorrelated* experience. We find ourselves simultaneously in the world of our own thoughts, and in a physical world that is not our own. Thus, although the 'points of view' of science and consciousness certainly intersect, neither can be reduced to the other. Bergson introduced this thesis, in his own manner, in terms of 'systems of images' in the famous opening section of *Matière et Mémoire*.[24] This requires more than is offered by either the scientistic approach, or phenomenological idealism. This is also the explanation of why we find Bergson's approach at the heart of lively debates in the contemporary French philosophical scene, ranging from Deleuze to Quentin Meillassoux, Pierre Montebello or Renaud Barbaras.

It is within the space of this impossible-to-fill critical distance that we find the location of the *artist*, as construed by Bergson. In other words, and borrowing from the fortuitous formulation of Jean-Marc Lévy-Leblond, *science is not art*. Just as positive metaphysics cannot be reduced to the task of revealing the ontological commitments of a scientific theory, the truly creative work of an artist does not consist in simply showcasing – by way of metaphor or simply illustration – a concept or hypothesis that one would find ready-made, and more precisely stated, in some scientific theory. De Vinci's paintings were not conceived to illustrate the golden ratio, similarly, Duchamp's installations were not intended to reveal to us the nature of the fourth dimension. It's quite the opposite. The artist is someone who is essentially 'distracted', Bergson tells us: he is a seer, not a scientist, he exists to reveal what scientists themselves cannot perceive, due to the very ontological commitments of their theories. In acknowledging the clairvoyance of artists we are not, of course, committing ourselves to a (slightly) suspect neo-romanticism. The artistic vision can have the most abstract and formal manifestations. Think of the work of François Morellet: although his artistic operations remain in the domain of the sensible, they are often indiscernible from a mathematical gestures. Art and science can also work together in an experimental mode, as shown by the 'constrained' fictions invented by Philippe Ramette, which illustrate so well the affinity between certain artistic practices, and the rigorous yet risky activities of science conceived as '*cosa experimentale*'.[25]

In any event, there is no *a priori convergence* between artistic creation and technoscientific research. It's rather their complementary divergence that we should be emphasizing, because it makes the *fiction* and *interpretation* that are inherent in art so valuable, even if there exist a wide range of potential interactions between art and science. These considerations suggest a re-evaluation of Bergson's notion of a 'fabulatory function'. This notion is originally associated with 'closed' morals and religion, but it's worth wondering if a fabulatory dimension isn't invariably a feature not just of religion and morality, but in the sciences as well. There are two reasons for this.

First, it is simply impossible to separate entirely science and culture. The two are irretrievably locked together. Lévy-Leblond was right when he observed that we find *interpreters* not only in music or theatre, but also in science, where they can be good or bad (in the latter case we tend to label them 'popularizers'). In today's highly complex sciences, the presence of an interpretive dimension is increasingly evident: the importance of the notion of a 'model' in contemporary science provides one illustration of this, another is linked to the fact that scientists are systematically confronted with the problem of 'underdetermination of theory by data' (a theme found in the philosophy of science, in Duhem, Quine and Atlan).

Next, as Dominique Janicaud has noted, it is necessary to distinguish between science itself and the power of science (Janicaud 1985). Somewhere in between, we find technoscientific research, which is less and less devoted to science and more and more devoted to its power. To put it another way, today's technoscientific research lies at the heart of a techno*economic* enterprise which is transforming nature at an industrial scale – and industrial speeds. As Hans Jonas has shown, what nature *is* now depends on what mankind can and wants to do with it. What we are dealing with here is not solely a matter of fact, but also of value, so the problem of the power of science isn't just a theoretical problem: it's an axiological problem, one which places us at the intersection of *knowledge* and *belief*.

However, to come back to Bergson's original intuition, confabulation goes hand-in-hand with closure, and it's to the latter notion that we need to attend if we are going to fully appreciate the critical requirement which underpins the relationship between philosophy – or culture at large – and science. The tension exerted by scientistic prejudices upon the formulation and popularization of scientific is an instance of the kind of closure that preoccupied Bergson, a relation which pushes the scientific world to close in on itself, rather than opening itself up to the cultural universe which in fact – paradoxically – feeds and nourishes it. In another sense, we can also wonder how to properly understand the distinction between *fiction* and *confabulation*. Here we venture a hypothesis: artistic fictions and inventions do not rest on the same *belief regime* as science. This becomes apparent when one examines the relationship between 'cognitive hallucinations' in science and the 'suspension of incredulity' which operates in fiction – and perhaps even more strongly in science fiction literature (and films). Blondlot's 'N-rays' constitute one classic example of a cognitive hallucination in the history of science, but we could just as easily cite controversies linked to the supposed discovery of the 'memory of water'. In such cases, how does one go about separating the wheat from the chaff? What distinguishes the rational belief that is an inherent component of good scientific practice from wholly unreal fictional creations? Is there any alternative to mounting a critique of (what we might call) the 'spontaneous ideology of scientists', or alternatively, a psychoanalytic investigation of their motivational and cognitive biases? What really differentiates the hypotheses of N-rays from that of the ether – a hypothesis that was revived on several occasions over the past two centuries? Or the memory of water from Ptolemy's epicycles, or the epigenetics of contemporary molecular biology? What is the difference between the 'psychical sciences' in which

Bergson was once seriously interested – prior to their fall into disrepute, and subsequent categorization as mere pseudo-sciences – and certain researches conducted today at the frontiers of art, anthropology and psychoanalysis? Such questions are on the horizon of our expanded Bergsonism: 'alterscience' may just be pseudo-science, but this should not blind us to the utopian potential of truly *open* science.

Envoi

Do we really need to spell it out? On all these questions, and on many others we have been unable to consider here, we are not proposing to scrutinize contemporary sciences – or their ordinary modes of practice – for *confirmations* of one or other of Bergson's doctrines or presentiments. Irrespective of whether we are dealing with the enactive theory of perception, the neurophysiological basis of memory, or any other line of contemporary research, intuitions do not exist to be rediscovered and confirmed. Their role, rather, is to raise new questions, and so in a fashion to make matters more complicated than they already are. Bergson succeeded in devising a method which makes it possible to discern and create new *problems* – the originality of some of the problems he formulated for himself still have the power to astonish us. It is in the light of the questions he bequeathed to us, and above all in the light of those he allowed us to pose in our turn – not any particular doctrine or vision of the world[26] – that we have to ask ourselves today, collectively, whether we still *want* to be Bergsonians, and whether we still *can* be.

> In brief, we possess even now a certain number of *lines of facts*, which do not take us as far as we want, but which we can extend hypothetically. I would like to follow some of these with you. Each, taken by itself, will lead us to a conclusion that is merely probable. But taking them all together, they will – by their convergence – bring before us such an accumulation of probabilities that we shall feel that we are on the road to certitude, or so I hope. Moreover, we shall come nearer and nearer to it through the joint effort of philosophers who will become partners. For, on this view, philosophy is no longer a construction, the systematic work of a single thinker. It needs, and unceasingly calls for, corrections and retouches. It progresses like positive science. And like a positive science, philosophy too is a work of collaboration.
>
> Bergson 1920: 7

The Parrhesia editorial team kindly granted permission to reproduce this article in this volume. We hereby express our gratitude.

Notes

1   Gilson (1959: 136). When writing this Gilson had a specific example in mind: 'The most overtly "Bergsonian"' of those who bear the mark of his influence, Édouard Le

Roy, has always refrained from adopting the same doctrines as the philosopher to whom he has rendered so many and so fervent tributes. [...] This wasn't a matter of betraying Bergson but rather of imitating him: whenever Bergson approached a new problem, he would always approach it as a wholly new problem, one requiring a wholly new effort.' On the import of the label 'Bergsonian', see Bianco (2015).

2   In this regard, the colloquium devoted to *Matière et Mémoire*, organized in Japan by Yasushi Hirai, Hisashi Fujita and Shin Abiko was exemplary ('The Anatomy of *Matter and Memory*: Bergson and Contemporary Theories of Perception, Mind and Time', 7th International Workshop of the Project Bergson in Japan, Tokyo and Kyoto, 10–13 December 2015). This manifesto – drafted between two sessions – is a direct result of this seminal event.

3   In the first chapter of *Matière et Mémoire* Bergson describes the phenomenon of 'total reflexion' which serves as an analogy for pure perception (2007c: 34).

4   See the introduction to 'Bergson et la science' (Miquel and Worms 2006).

5   A 1938 interview with Bergson, cited by Adolphe (1951: 4).

6   'Recoupement', as Bergson uses the term, involves both intersecting and double-checking.

7   Henri Bergson, 'Discussion à la Société française de philosophie, 28 novembre 1907', in 2011: 353.

8   Henri Bergson, 'Discussion à la Société française de philosophie, 28 novembre 1907', in 2011: 353.

9   The 'tightening' of problems is complementary to the process of '*recoupement*' (see note 8 above). The best example of the strategy is provided by *Matter and Memory*, and the decision to concentrate on the phenomenon of memory, and more specifically, the exemplary case of aphasia, which allows Bergson to broach the question of mind-body dualism 'at the point where the activity of matter brushes against that of spirit'.

10 A point Camille Riquier brings out well in his *Archéologie de Bergson* (2007: 253).

11 Cf. Bergson (2007b: 239): 'The dialectic is necessary to put intuition to the test, but also so intuition can be refracted in concepts, and propagated to other people.'

12 A note from 16 May 1912, concerning Joseph Deseymard's *La Pensée d'Henri Bergson*, which was published in November 1912 (Fonds Doucet, BGN 2966), and cited by Riquier (2007: 192).

13 Although it's clear that Bergson didn't envisage philosophy being able to contribute directly to the development of new scientific theories by substituting metaphysical considerations for scientific explanation, it's equally clear that he didn't see its role as being limited to one of epistemological clarification. In certain cases, a philosopher can propose to arbitrate between competing theories, based on their ability to account for 'experience in its entirety', or at least to contribute to an inquiry aiming at a more comprehensive view of reality: it's in this way that Bergson was an 'interventionist' in the field of biology, to use Jean Gayon's expression. We have already cited this remarkable passage: 'The *intuition* that I have spoken of can only enter into play after one has studied, deepened, and often even expanded upon everything that empirical science has to say on a certain issue.' By following the lines of facts, philosophers can sometimes uncover new ones.

14 Henri Bergson, 'Le parallélisme psycho-physique et la métaphysique positive', in Bergson (2011: 249).

15 Worms (2004). This theme also resonates in a book inspired by the work of Winnicott and de Bowlby (2013).

16 On this topic, see the remarkable analysis provided by Pierre Montebello (2003).
17 'Being possesses a transductive unity, i.e. it can dephase in relation to itself, and overflow itself from either side of *its centre*' (Simondon 2005: 31).
18 By way of example, this is the approach adopted by Barry Dainton in his studies devoted to temporal experience. Yasushi Hirai provided a very interesting examination of Dainton's work in a lecture at the 'The Anatomy of Matter and Memory' conference in December 2015.
19 See the works of David Lapoujade, Hisashi Fujita or Sébastien Miravète.
20 See Elie During's chapter in this book.
21 There is no shortage of textual evidence of this thesis, but for present purposes three occurrences will be sufficient. On p. 82 of *Durée et Simultanéité* (Bergson 2009), we find 'What is real is what is measured by a real physicist'. On p. 207: 'Real time is ... the time that physicists perceive and measure ....' Lastly, on p. 209: 'Real time, measured by real clocks ...'.
22 Regarding the context of Bergson's confrontation with Einstein, see Canales 2015, as well as the critical edition of *Durée et simultanée* by Elie During (Bergson 2009).
23 In this respect Bachelard was obviously not wrong to point to Bergson's geometrical obsessions, and to criticize his lack of interest in the problems posed by the use of probabilities in science, and more generally algebra. That did not prevent his praising the creative syntheses achieved by the new mechanics in very Bergsonian terms, such as 'spiritual élan' and even 'élan vital' (Bachelard 1934: 183).
24 'No philosophical doctrine disputes the fact that the same images can enter at the same time into two distinct systems, one belonging to science, wherein each image, related only to itself, possesses an absolute value; and the other, the world of consciousness, wherein all the images depend on a central image, our body, the variations of which they follow' (Bergson 2007c: 21).
25 We take this expression, and the two examples, from Jean-Marc Lévy-Leblond.
26 Is the universe itself in the process of becoming? It will remain a matter of taste as long as the reasons for holding one view or the other have not been fully spelled out. Following William James, we need to make sure that genuine becoming makes a real difference. This is our advice to those who enjoy the sparring between 'presentists' and 'eternalists'.

References

Adolphe, L. (1951), *La Dialectique des images*, Paris: Presses universitaires de France.
Bachelard, G. (1934), *Le Nouvel esprit scientifique*, Paris: Presses universitaires de France.
Bailly, F. and G. Longo (2006), *Mathématiques et sciences de la nature: la singularité physique du vivant*, Paris: Hermann.
Bergson, H. (1920), *Mind-Energy*, trans. H. Wildon Carr, New York: Henry Holt and Company.
Bergson, H. (2007a), *La Pensée et le mouvant*, Paris: Presses universitaires de France.
Bergson, H. (2007b), *L'Évolution créatrice*, Paris: Presses universitaires de France.
Bergson, H. (2007c), *Matière et Mémoire*, Paris: Presses universitaires de France.
Bergson, H. (2009), *Durée et simultanéité: à propos de la théorie d'Einstein,* Paris: Presses universitaires de France.
Bergson, H. (2011), *Écrits*, Paris: Presses universitaires de France.

Bianco, G. (2015), *Après Bergson: portrait de groupe avec philosophe*, Paris: Presses universitaires de France, coll. Philosophie française contemporaine.

Canales, J. (2015), *The Physicist and the Philosopher: Einstein, Bergson, and the Debate that Changed our Understanding of Time*, Princeton: Princeton University Press.

Gilson, É. (1959), 'Souvenir de Bergson', *Revue de métaphysique et de morale*, 64: 136.

Janicaud, D. (1985), *La puissance du rationnel*, Paris: Gallimard.

Longo, G. and M. Montévil (2014), *Perspectives on Organisms: Biological Time, Symmetries and Singularities*, Berlin: Springer.

Miquel, P.-A. and F. Worms, eds, *Annales Bergsoniennes*, no. 3, Paris: Presses universitaires de France, coll. Épiméthée.

Montebello, P. (2003), *L'autre métaphysique: essai sur la philosophie de la nature, Ravaisson, Tarde, Nietzsche et Bergson*, Paris: Desclée de Brouwer.

Rancière, J. (2004), *The Flesh of Words: The Politics of Writing*, trans. C. Mandell, Redwood City: Stanford University Press.

Riquier, C. (2007), '"Voir et cependant ne pas croire": Intuition et méthode chez Bergson', *Transparaître*, no. 1, 'L'intuition'.

Riquier, C. (2009), *Archéologie de Bergson*, Paris: Presses universitaires de France.

Ruyer, R. (2012), *Néo-finalisme*, Paris: Presses universitaires de France.

Simondon, G. (2005), *L'individuation à la lumière des notions de forme et d'information*, Grénoble: Jérôme Millon.

Worms, F. (2004), *Bergson ou les deux sens de la vie,* Paris: Presses universitaires de France.

Worms, F. (2013), *La Vie qui unit et qui sépare*, Paris: Payot.

Index

A-series (McTaggart) 16
A-time/B-time 186, 232
Abgrund 54–5
Achilles 91
actual/virtual 15–18
affordance 100, 105, 133–5
aging 93, 188
Alexandroff interval 190
amnesia 36–7, 49
The Analysis of Matter (Russell) 143
The Analysis of Mind (Russell) 142–3
anamnesis 58–9
animadversion 63, 75n.1
antinomies theory 67, 70
aphasia 28, 34, 37, 60
Arecchi, F. T. 206
Aristotle 177, 223
artificial intelligence 3, 8n.3, 197, 209
associationist approach 15
Aufhebung 64, 68
'automatic conservation of past' thesis 127
automatic recognition 119

B-series (McTaggart) 16
Baars, B. J. 198
Bachelard, Gaston 55, 236, 241n.23
Bailly, F. 230
Barbour, Julian 191
Barnard, W. 6, 142, 145
Barthélemy-Madaule, Madeleine 72
Bassui 94
Bastian, Henry Charton 29, 32–3
Bayesian inference 195, 204–9
behaviourism 108, 114n.3
Benjamin, Walter 54, 58–9
Bergson, Complexity, and Creative Emergence (Kreps) 6
Berlin Wall 154
Bernard, Claude 21–2
Bertalanffy, Ludwig von 122
Berthelot, R. 16

blindsight patients 118
Block, Ned 4
Blondlot's 'N-rays' 238
Bohm, David 84
Bohr, Niels 122
Boltzmann, L. 122
brain
  and Bayesian inference 206
  Bergson's views 195–9
  as a concrete wave 90–1
  and contemporary selectionist
    theories 99
  and external world 82
  and habit-memory 100
  as a hologram 84–6
  introduction 2–3
  and Libet's experiment 47–8
  material in the universe 19
  and memory 44–5, 48–9, 53, 97, 104–5, 134, 209
  and motor schema 109–10, 114nn.4-5
  neuron-system 56
  and perception 108
  storehouse of localized functions 28, 33–8, 40
  and visual perception of object 124
Brentano, Franz 147–8
Broca, Paul 28, 35, 39n.8
Bruner, Jerome 107–8

Canguilhem, Georges 26, 231
Cartesian dualism 43–4
Cattell, J. M. 27
causation 46–7, 50, 171
cause-effect loop 203, 209
cerebral cortex 28
Chalmers, David 81, 140, 144
Changeux, Jean-Pierre 53–4, 99
Chen, I. 108
Chrétien, Jean-Louis 71
Cinema (Deleuze) 133

Cinematic model 146–7, 183, 185–7, 192n.6
clôture 60
coexistence 16, 18–19, 113, 175–92
cogito 55
cognitive psychology 107–13, 113n.1
colour 44, 49, 111–12, 123, 126, 140, 142, 144–5, 170, 198, 214
conceptualism 195, 204
concrete extension 20
condensation 214, 218
The Conscious Mind (Chalmers) 140
consciousness
   and abstract computations 90
   choice or discernment 99
   diverse range of contents in 152
   and dualism 45–6, 48
   and elementary physical particles 142, 144
   excluding the past 155–6
   extensionalist view of 126
   and Holt 141
   and Husserl 229
   intersection with science 237
   introduction 4
   and Libet's experiment 47
   and material time units 111
   and memory 48, 151–2, 155–6
   not a light projector 110
   passive 195–7
   the past of 18
   and the physical world 140
   a process 16–19
   retains image of situations 166
   spiritual dimension of 20–2
   stream of 146–7, 149, 152–3, 178, 220
   temporal 214
   and unconsciousness 199–200
   the world as a moving continuity 172
constructivist approach 3, 8n.10, 237
contemporaneity 19, 188–90
contraction 126, 213–22, 215n.4, 221–2n.9
cosmology 43, 50n.2, 142
   see also space, universe
Creative Evolution (Bergson) 20, 63, 69–70, 134, 176, 180–1, 183, 185, 229
Creative Mind (Bergson) 17, 72
criticism 72
Critique of Judgement (Kant) 73

Critique of Practical Reason (Kant) 73
Critique of Pure Reason (Kant) 65–75, 175, 177–8

Dainton, Barry 4–7, 9n.26, 149, 219–20
date 163–9, 171–3
De Vinci, Leonardo 237
death 58, 170, 231
dédoublement (dual personality) 37
Dehaene, S. 99, 198–9
déjà vu 200–4
Deleuze, Gilles 38, 63, 71–2, 76n.8, 133–4, 176–7, 224–5, 227
déracinement 59
Derrida, Jacques 55, 58, 64
Descartes, René 44, 46, 54, 73–4, 80–1, 141–2, 180–1
détente 182
determinism 181–2, 224
direct perception theory 120–2, 133
direct realism 81, 142, 213, 219
direct recollection theory 127, 130–1nn.23, 27–8
dual perception system 117–29, 129nn.4–6
dual personality 37
dualism
   arguments against interactionist 46–7
   conclusion 50
   desire to avoid 140–1
   and image space 209
   interactionist 45–6
   introduction 43
   and Libet's experiment 47–8
   of matter and spirit 208
   and memory 48–9
   and mind 43–5, 50nn.3–4
   and sensory qualities in the external world 144–5, 157n.9
Dummett, Michael 139
duration 16–17, 58, 152, 177, 185, 188, 195, 199, 232–3
Duration and Simultaneity (Bergson) 175–7, 187, 233
durée 145, 153, 223–4, 229–30, 232–5
'dwarfs in a brain' 197
dynamic motion 147

Ebbinghaus, Hermann 26
Ebbinghaus illusion 118

echolalia 28–9, 39nn.8
Eddington, Arthur 180
Edelman, G. M. 99
Ego 19, 67
Einstein, Albert 175, 178, 233–5
élan vital 60
emergentism 46, 228–30
empirical metaphysics 71–2
empirical psychology 74
engram 49
entropy 22
epiphenomenalism 45, 47
eternalism 153–4, 180–1, 231–2
Evangel 60
event-individuals 170–2
events
    arbitrariness of individuation of 172
    description/category of 169–70, 173, 174nn.4–5
    introduction 16–17
    the primary realities 102
    type 171
    unique 58
excluded middle principle 230
Exner, S. 214
Exner's experiment 216
Extensional model 148–52, 216–19, 221nn.7-8
extensionalism 213–22
externalism 85

fabulation 57–8, 60
'fabulatory function' 237–8
Fast and Slow (Kahneman) 3
Fechner, Gustav 26
Fichte, Johann Gottlieb 67
Forest, D. 109
Foucault, Michel 26
frame-time 186, 189–90
free energy minimization principle 205
Freud, Sigmund 29, 34–5, 40nn.11, 14–15
Friston, K. J. 205

Gabor, Dennis 82
Ganong's Review of Medical Physiology 120
Gazzaniga, M. S. 3
General Systems Theory (Bertalanffy) 122
Gestalt theory 108, 113–14n.2
Gibson, J. J. 81, 86, 97–9, 102–5, 133, 195–6

Gibsonian cubes 87
Gilson, É. 223, 239–40n.1
gist 120, 129n.7
global workspace theory 198, 205
God 4, 67, 72
Gödel, Kurt 236
Goldscheider, A. 27
Goldstein, Kurt 28, 56
Goodale, M. 117
Goodman, C. C. 108
growing block model 154, 156

Hacking, Ian 25–6, 35, 38, 39nn.1–2, 41n.26
Halbwachs, Maurice 38
'hard problem' (Chalmers) 81
Hardcastle, Valerie 82
Hegel, Georg 64, 67, 81
Heidegger, Martin 21, 54, 57, 73, 183
Heracliteanism 186
Hesiod 58
heterogeneity 123–7, 165, 226
Høffding, Harald 145
holism 32, 37
holography 82–90, 133–4
Holt, E. B. 105, 141
homogeneity 165, 235
'How Did We Get Here from There?' (Williamson) 139–40
Hume, David 47
Husserl, Edmund 5, 54, 59, 147–8, 229

idealism 81, 98, 122, 143, 237
identity of indiscernibles principle 182
image space 208–9
'image-body' 56
'image-or scheme-body' 55–6
immanence 229–30
Inaga, M. 6
incompleteness theorems (Gödel) 236
integrated information theory 3, 8n.12
'Intellectual Effort' (Bergson) 173
intelligence 22, 57, 65, 69–71, 225, 227, 233
intuition 22, 65, 68, 226–7, 240n.9
invariance structure 89–90
The Invention of Memory (Rosenfield) 28
inverse Bayesian inference 195, 204–9

Jackson, John Hughlings 34
James, William 5, 45, 91, 105, 141, 143–4, 147, 149, 183, 214, 220
Janicaud, Dominique 238
Jonas, Hans 238

Kahneman, Daniel 3
Kaku, Michio 3–4
Kant, Immanuel 55–7, 63–74, 75–6nn.5–7, 81, 110–11, 148, 175–80, 187
kerygma 60
Koch, C. 197
Kreps, David 6
Kullback–Leibler divergence 205
Külpe, O. 27
Kussmaul, Adolf 29
Kyoto Manifesto 7, 223–39

La philosophie de la volonté (Ricoeur) 54
Lacan, Jacques 71
Lange, N. N. 27
Langevin, Paul 178, 187–8, 193n.7, 234–5
Langevin's paradox 188, 234
Laplace, Pierre-Simon 181–3, 182n.2
Leibniz, Gottfried 175, 182
L'Energie Spirituelle (Bergson) 149
Lenin, Vladimir 226
Levinas, Emmanuel 60
Levine, R. 108
Lévy-Leblond, Jean-Marc 237–8
Libet, Benjamin 47–8, 50, 197
Lichtheim, Ludwig 28–9
life 21–2, 45, 50, 57–9, 149, 155, 182, 227, 229–31
localization theory 35
the localization theory of the memories (Wernicke) 28
Locke, John 141–2
Longo, Guiseppe 220, 237
loop quantum gravity theory 4
Lynds, P. 92

McTaggart, John 4, 23
Maine de Biran 54
Malabou, Catherine 54
Marcel, Gabriel 56
Marion, Jean-Luc 74
Marx, Karl 57, 224
Massimini, M. 3

Matter and Memory (Bergson)
 and aphasia 60
 and artificial intelligence 209
 conceiving the physical 141–3
 conclusions 38–9, 42nn.32–4
 consciousness and the physical world 139
 and constitutive role of memory 182–3
 construction of theories in 25–6
 and Deleuze 133
 and direct perception/memory theory 128
 dual perception doctrine 119–22
 and dualism 43, 144
 far removed from modern science 195
 and Frédéric Worms 63
 and image 82
 introduction 1, 4–5, 7, 8nn.1, 8
 and Kant 55–6, 74, 76n.11, 110
 localization of memories 28–9, 32–5, 40n.13
 memory impacting on consciousness 151
 and Merleau-Ponty 97
 and 'my present' thesis 126
 and past psychological states 155
 and perception of time 214, 221n.3
 psychodynamics of Ribot 35–8, 41nn.27–8
 relation between conceptualism and nominalism 204
 relation between present and past 181
 succession of experiences 220
 systems of images 237, 241n.24
 theory of cognition 70, 75n.6
 and *Time and Free Will* 59
 and *Two Sources* 53–4, 57
 unconscious memories 183–4
 unity of real time 177
 and virtual/actual 15, 17
Maturana, H. R. 22
measurement critique 233
memory
 acquired 164
 and actual/virtual duality 15–19
 arrangement in a line 161–74
 Bergson explanation of 123–4
 and the brain 44–5, 48–9, 53, 97, 104–5, 134, 209

connection and disconnection
    195–209
and consciousness 48, 151–2, 155–6
and distinct perceptions 120
dualism 48–9
and history 53–60
and image space 209
imports the past into the present 126
localization of 28–9, 32–5
motor 47
not in the brain 110, 114nn.4-5
ontological status in the universe
    104–5
and order 167
and the past continuing 181–2
and perception 27–8, 101–2, 201, 204,
    208
primary 91
pure 185
and recollection theory 120
sciences of 25, 38
and sleeping 20
spontaneous 164
through the brain 134
two types of 99–102
Memory, History, Forgetting (Ricoeur) 54,
    57–8
'Memory of the Present and False
    Recollection' (Bergson) 135, 183
'Mental Presence-Time' (Stern) 149
Merleau-Ponty, Maurice 53, 56, 97
Metaphysics of Morals (Kant) 64
Meynert, Theodor 28, 40n.12
Milner, D. 131
mind-body relation 1, 47
Minkowski spacetime 232, 234, 236
Minturn, A. L. 107
mnemosyne *see* memory
Monism 144
Montévil, M. 230
Morellet, François 237
motion 2, 8, 87–8, 90–2
motor diagram 28–9, 35–6
motor memory 47
motor schema 109–11

motor speech centre 28
Müller, Georg Elias 26
multiple constitution principle 127

multiplicity
    numerical 164–5, 169
    qualitative 164–5, 167, 169
Münsterberg, Hugo 27
Murphy, G. 108
'Mussati's illusion' 88

Naccache, L. 198
nebula/nebulalization 59–60
Neuronal Man: The Biology of the Mind
    (Changeux) 53
neurons 49, 53, 56, 197–8, 205
neutral monism 141
'New Realists' 141–2, 145, 157n.10
Newton, Isaac 178
Nietzsche, Friedrich 58–9
nominalism 195, 204
non-contradiction principle 230
normopathe 60
Nottale, L. 92

On Aphasia (Freud) 29, 34
'On Some Motifs of Baudelaire'
    (Benjamin) 54, 58
'order' concept 165–7
oscillation 89, 94, 204
otherness 17–18, 231
overlap model (Dainton) 219

Panero, Alain 64
panpsychism 46, 93–4, 213, 219
panqualityism 144
parallelism 35, 69
paralogisms 67
Parrhesia (journal) 7
Pascal, Blaise 180
passive consciousness hypothesis 195
past, the
    automatic conservation of 162,
        166–8
    contemporaneous formation with the
        present 161–2
    continuance of 92n.1, 180–1
    and déjà vu 202–4
    existence of 183–5, 192n.3
    trans-temporality of 162–3
    unrepeatability of 168–9
Pattie, Howard 122
Péguy, Charles 73

perception
    and association 35–7
    attentive 25
    Bergsonian theory of 98–9
    Bergson's dual theory 119–20
    Bruner's cognitive theory of 107–8
    of change 149, 152, 155
    connection and disconnection 195–209
    and consciousness 141
    and definition of philosophy 108–13
    and depth 102–4
    direct realist view of 156
    dual perception system 117–29
    and dualism 43–6, 49
    and embodiment 79, 85, 89–90, 92, 94
    and Gibson 133–5
    human 107
    introduction 1–2, 15
    and memory 53, 57, 97–8, 102–5, 161–8, 182, 184
    natural 186
    pure 18–21
'The Perception of Change' (Bergson) 65, 152, 155
perception theory (Bergson) 98–9, 102, 108–13
Petit, Jean-Luc 54
phenomenal unity 152
phenomenology 2, 4, 54, 224, 227
philosophy 107–13, 228, 232, 236, 240n.13
The Philosophy of Will (Ricoeur) 55
Physics (Aristotle) 177
Piaget, Jean 191, 234
Pittenger, J. B. 95
'plane of immanence' 229
Poincaré, Jules 233
Popper, K. 45–7
present, the
    absolute 182, 192n.2
    contemporaneous formation with the past 161–2
    and déjà vu 202–4
    leaks from the past 181
    too much emphasis on 183
presentism 154–6, 180–1, 231–2
Pribram, Karl 84
primal impressions 151–2
priming images 199

Principles of Psychology (James) 144, 146–7
Proust, Marcel 38
psycho-cerebral identity 45
psychodynamics 35–8
psychophysics 108
Ptolemy's epicycles 238
Putnam, Hilary 140

qualia
    and body in image space 208–9
    colour 126
    introduction 4
    origin of 81–2
    and time-scaling 92
qualitative individuality 169
Que faire de notre cerveau? (Malabou) 54
Quine, W. V. 140

Ramette, Philippe 237
Rancière, Jacques 225
'Rebooting Panpsychism' 7, 9n.27
recollection 200–5
reductionism 215, 219, 228–30
relativism 72
The Relativistic Brain (Cicurel/Nicolelis) 90, 92
relativity 72, 188, 235–6
Renouvier, Charles 59, 112–13
resemblance concept 112–13
retentionalism 147–9, 151, 153, 157n.18, 213, 220–1
Rewriting the Soul: Multiple Personality and the Sciences of Memory (Hacking) 25
Ribot, Théodule 36–8, 40nn.19–24, 109
Ribot's law 36
Ricoeur, Paul 53–60
Riggio, A. 5
Riquier, Camille 63–4, 67–75, 76n.10
Rosenfield, Israel 28, 39n.7
Rovelli, Carlo 191
Russell, Bertrand 141, 142–4, 157nn.6–7
'Russellian Monism' 141
Ruyer, R. 229

schème/schéma 70
Schilder, Paul 56
Schopenhauer, Arthur 67, 70

science is not art (Lévy-Leblond) 237
Scientific Metaphysics (Ross/Ladyman/
    Kincaid) 5
Scientific Revolution 140–1
Searle, John 81
*Seeing Things as They Are: A Theory of
    Perception* (Searle) 81
Sein und Zeit (Heidegger) 54, 57
'sensorimotor' mechanism 126, 130n.19
Shaw, R. E. 95
Simondon, G. 230–1
simultaneity 175–6, 187–9, 233
Smith, William George 26, 39n.4
Smolin, Lee 4
Sommer, Robert 34, 39–40n.10
soul, the, 'scientized' and secularized 25
space 165–6, 169, 172–6, 189, 191, 198–9,
    208–9, 232–3, 235
spacetime 178–9, 187–9, 191, 235
Spinoza, Baruch 74, 81
Stern, William 149
Strawson, Galen 140
Stream of Consciousness (Dainton) 4–5
succession 165–6, 213–14
Supplementary Motor Area (SMA) 47–8
'symbol-body' 56
synthesis 214–16
systems theoretic approach 121–2

The Taming of Chance (Hacking) 38
'Temporal Consciousness' (Dainton) 146
temporal consciousness theory (Bergson)
    219–21
temporal experience
    and consciousness 153, 157n.17
    Extensional model 156
    introduction 5
    and time 153–6
Theogony (Hesiod) 58
thing-individuals 170–2
things 37, 45, 65–6, 70–2, 103, 120–2, 127,
    141, 143, 170
'The Three Criticisms' (Kant) 64
time
    Bergson's idea of 149
    cannot be measured 124, 129n.15
    and duration 16, 175–7
    ecstatic dimensions of 18
    and events 102

flowing 4
form of 177–80
'hybrid' growing universe theory of
    127
linear 167–8
'lived' and 'proper' 235
not linear and divisible 102–3, 164–5
perceptive/material 111–13, 114n.7
philosophy of 4
'real time' from 'fictitious' 187, 189
relational 190–2
scales and invariance laws 93
'succession' and 'order' 165–6
and temporal experience 90–3, 153–6
and twins paradox 234
Time and Free Will (Bergson) 59, 69, 73,
    112, 164, 176
Time Reborn (Smolin) 4
Tononi, G. 3, 211
Tractatus (Wittgenstein) 178
transcendental analytic 70–3
'transcendental empiricism' (Deleuze)
    134
transcendental judgements 65–7
transcendental psychology 72
transcendental schematism 75, 76n.12
Tulving, Endel 49
Turing machines 90
twins paradox 187, 188–91, 193nn.7, 9,
    233–4
'two orders' 230
Two Sources of the Moral and the Religion
    (Bergson) 53, 56–7, 60, 73, 134

'Unconscious Simulation in the
    Hypnotic State' (Bergson) 37,
    41nn.29–31
unconsciousness 199–200
Unfashionable Observations (Ricoeur)
    58
universe 92, 101, 127, 153, 175–7, 209, 233,
    239, 241n.26
'The Unreality of Time' (McTaggart) 4

Varela, F. H. 22
virtual 15–23, 101–5, 126–9, 176, 184, 199,
    208, 230, 232
The Voluntary and the Involuntary
    (Ricoeur) 54

Weil, Simone 59
Weizsäcker, Viktor von 56
Wernicke, Carl 28, 34–5
Whitehead, Alfred 178, 183, 187, 190
Wholeness and the Implicate Order (Bohm) 84
Williamson, Timothy 140
Wittgenstein, L. 140, 177–8

Wolffian division 72
Worms, Frédéric 63, 229
Wundt, Wilhelm 27, 37, 109

Young, J. Z. 99

Zeno's paradox 91, 135
zombies 197

www.ingramcontent.com/pod-product-compliance
Lightning Source LLC
Chambersburg PA
CBHW071817300426
44116CB00009B/1345